On Extended
Hardy-Hilbert Integral
Inequalities
and Applications

On Extended Hardy-Hilbert Integral Inequalities and Applications

Bicheng Yang

Guangdong University of Education, China

Michael Th Rassias

Hellenic Military Academy, Greece

NEW JERSEY · LONDON · SINGAPORE · BEIJING · SHANGHAI · HONG KONG · TAIPEI · CHENNAI · TOKYO

Published by

World Scientific Publishing Co. Pte. Ltd.

5 Toh Tuck Link, Singapore 596224

USA office: 27 Warren Street, Suite 401-402, Hackensack, NJ 07601

UK office: 57 Shelton Street, Covent Garden, London WC2H 9HE

Library of Congress Cataloging-in-Publication Data
Names: Yang, Bicheng, author. | Rassias, Michael Th., 1987–   author.
Title: On extended Hardy-Hilbert integral inequalities and applications /
    Bicheng Yang, Guangdong University of Education, China,
    Michael Th Rassias, Hellenic Military Academy, Greece.
Description: New Jersey : World Scientific, [2023] | Includes bibliographical references and index.
Identifiers: LCCN 2022051280 | ISBN 9789811267093 (hardcover) |
    ISBN 9789811267109 (ebook for institutions) | ISBN 9789811267116 (ebook for individuals)
Subjects: LCSH: Inequalities (Mathematics) | Mathematical analysis.
Classification: LCC QA295 .Y37 2023 | DDC 515/.46--dc23/eng20230117
LC record available at https://lccn.loc.gov/2022051280

British Library Cataloguing-in-Publication Data
A catalogue record for this book is available from the British Library.

For any available supplementary material, please visit
https://www.worldscientific.com/worldscibooks/10.1142/13164#t=suppl

Desk Editors: Logeshwaran Arumugam/Nijia Liu

Typeset by Stallion Press
Email: enquiries@stallionpress.com

Printed in Singapore

Preface

Hilbert-type inequalities, including Hilbert's inequalities proved in 1908, Hardy–Hilbert-type inequalities proved in 1934, and Yang–Hilbert-type inequalities first proved around 1998, play an important role in analysis and its applications. These inequalities are mainly divided into three classes: integral, discrete, and half-discrete. During the past 20 years, there have been many research advances on Hilbert-type inequalities, especially on Yang–Hilbert-type inequalities.

In the current monograph, applying weight functions, the idea of parameterization, as well as techniques of real analysis and functional analysis, we prove some new Hilbert-type integral inequalities as well as their reverses with parameters. These inequalities constitute extensions of the well-known Hardy–Hilbert integral inequality. The equivalent forms and some equivalent statements of the best possible constant factors associated with several parameters are considered. Furthermore, we also obtain the operator expressions with the norm and some particular inequalities involving the Riemann zeta and Hurwitz zeta functions. In the form of applications, by means of the beta and gamma functions, we use the extended Hardy–Hilbert integral inequalities to consider several Hilbert-type integral inequalities involving derivative functions and upper-limit functions. In the last chapter, we consider the case of Hardy-type integral inequalities. The lemmas and theorems within provide an extensive account of these kinds of integral inequalities and operators.

The monograph comprises six chapters. In Chapter 1, we introduce some recent developments in Hilbert-type integral, discrete, and half-discrete inequalities. In Chapter 2, using weight functions, a new Hilbert-type integral inequality with a nonhomogeneous kernel is established, and the case of the homogeneous kernel is deduced. The equivalent forms as well as some equivalent statements of the best possible constant factors associated with several parameters are obtained. We also consider the operator expressions, some applications, and the reverses. In Chapters 3–5, applying the extended Hardy–Hilbert integral inequalities and using the beta and gamma functions, we study in a step-by-step manner some new kinds of Hilbert-type integral inequalities involving derivative functions and upper-limit functions. In Chapter 6, we consider the case of Hardy-type integral inequalities with their operator expressions.

An effort has been made for this monograph to hopefully be especially useful to graduate students of mathematics, physics, and engineering, as well as researchers in these domains.

Bicheng Yang
Department of Mathematics
Guangdong University of Education,
Guangzhou, P. R. China
bcyang@gdei.edu.cn;
bcyang818@163.com

Michael Th. Rassias
Department of Mathematics and Engineering Sciences
Hellenic Military Academy,
Greece & Institute for Advanced Study
Program in Interdisciplinary Studies
Princeton, New Jersey, USA
mthrassias@yahoo.com;
michail.rassias@math.uzh.ch

About the Author

Michael Th. Rassias is an Associate Professor at the Department of Mathematics and Engineering Sciences of the Hellenic Military Academy as well as a visiting researcher at the Institute for Advanced Study, Princeton. He obtained his PhD in Mathematics from ETH Zürich in 2014. During the academic year 2014–2015, he was a postdoctoral researcher at the Department of Mathematics at Princeton University and the Department of Mathematics at ETH Zürich, conducting research at Princeton. While at Princeton, he co-edited with John F. Nash, Jr., the volume *Open Problems in Mathematics*, Springer, 2016. He has received several awards in mathematical problem-solving competitions, including a silver medal at the International Mathematical Olympiad of 2003 in Tokyo. His current research interests lie in mathematical analysis, analytic number theory, and, more specifically, the Riemann hypothesis, Goldbach's conjecture, the distribution of prime numbers, approximation theory, functional equations, and analytic inequalities.

Bicheng Yang was born in Shanwei, Guangdong, in China, on August 18, 1946. He currently works in the Department of Mathematics at the Guangdong University of Education, China. He obtained a BSc in mathematics from South China Normal University in 1982. His current research interests include analytic inequalities; extensions of Hilbert's inequality with best constant factors and applications; and extensions of weight inequalities and applications, especially Hardy–Hilbert-type inequalities, as well as sequences, series, summability, and improvements to the Euler–Maclaurin's summation formula and applications. He is currently working on a project supported by the National Natural Science Foundation of China (No. 61772140), which will be completed in 2022. He has published 524 papers in international journals. His publications also include 12 edited books and 15 book chapters in Springer volumes.

Acknowledgments

B. C. Yang: This work is supported by the National Natural Science Foundation (No. 61772140), and the Science and Technology Planning Project Item of Guangzhou City (No. 201707010229). We are grateful for this help.

Contents

Chapter 1

Introduction

1.1 Hilbert's, Hardy–Hilbert's, and Hardy–Hilbert-Type Inequalities

Suppose that $f(x), g(y) \geq 0$ $(x, y \in \mathbf{R}_+ = (0, \infty))$,

$$f, g \in L^2(\mathbf{R}_+) = \left\{ f; ||f||_2 = \left(\int_0^\infty |f(x)|^2 dx \right)^{\frac{1}{2}} < \infty \right\},$$

$||f||_2, ||g||_2 > 0$. The following Hilbert integral inequality and its equivalent form (cf. Ref. [1]) hold true:

$$\int_0^\infty \int_0^\infty \frac{f(x)g(y)}{x+y} dx \, dy < \pi ||f||_2 ||g||_2, \qquad (1.1)$$

$$\left[\int_0^\infty \left(\int_0^\infty \frac{f(x)}{x+y} dx \right)^2 dy \right]^{\frac{1}{2}} < \pi ||f||_2, \qquad (1.2)$$

where the constant factor π is the best possible.

Assuming that $a_m, b_n \geq 0$ $(m, n \in \mathbf{N} = \{1, 2, \ldots\})$,

$$a = \{a_m\}_{m=1}^\infty \in l^2 = \left\{ a; ||a||_2 = \left(\sum_{m=1}^\infty |a_m|^2 \right)^{\frac{1}{2}} < \infty \right\},$$

$$b = \{b_n\}_{n=1}^\infty \in l^2,$$

$||a||_2, ||b||_2 > 0$, we still have the following discrete Hilbert's inequality and its equivalent form:

$$\sum_{n=1}^{\infty} \sum_{m=1}^{\infty} \frac{a_m b_n}{m+n} < \pi ||a||_2 ||b||_2, \qquad (1.3)$$

$$\left[\sum_{n=1}^{\infty} \left(\sum_{m=1}^{\infty} \frac{a_m}{m+n} \right)^2 \right]^{\frac{1}{2}} < \pi ||a||_2, \qquad (1.4)$$

with π being the best possible constant factor again (cf. Ref. [6]).

Using inequality (1.2), we may define a Hilbert integral operator

$$T : L^2(\mathbf{R}_+) \to L^2(\mathbf{R}_+)$$

as follows (cf. Ref. [2]).

For any $f \in L^2(\mathbf{R}_+)$, there exists a $Tf \in L^2(\mathbf{R}_+)$ satisfying

$$Tf(y) = \int_0^\infty \frac{f(x)}{x+y} dx \, (y \in \mathbf{R}_+).$$

By (1.2), it follows that

$$||Tf||_2 < \pi ||f||_2.$$

Then, T is bounded by

$$||T|| := \sup_{f(\neq \theta) \in L^2(\mathbf{R}_+)} \frac{||Tf||_2}{||f||_2} \leq \pi.$$

Since the constant factor π in (1.2) is the best possible, we have

$$||T|| = \pi.$$

In view of (1.4), we may define a Hilbert operator $T : l^2 \to l^2$ as follows (cf. Ref. [3]).

For any $a = \{a_m\}_{m=1}^{\infty} \in l^2$, there exists a $Ta \in l^2$ satisfying

$$Ta(n) = \sum_{m=1}^{\infty} \frac{a_m}{m+n} \ (n \in \mathbf{N}).$$

Then, by (1.4), we have

$$\|Ta\|_2 \leq \pi \|a\|_2.$$

Then, T is bounded by

$$\|T\| := \sup_{a(\neq\theta)\in l^2} \frac{\|Ta\|_2}{\|a\|_2} \leq \pi.$$

Since the constant factor π in (1.4) is the best possible, we have

$$\|T\| = \pi.$$

In 2002, Zhang [4] studied the above two operators and established some new improvements in (1.1) and (1.3).

In 1925, by introducing a pair of conjugate exponents (p, q) $(\frac{1}{p} + \frac{1}{q} = 1)$, Hardy [5] presented the following extensions of (1.1)–(1.4).

For $p > 1, f(x), g(y) \geq 0$,

$$f \in L^p(\mathbf{R}_+) = \left\{ f; \|f\|_p = \left(\int_0^{\infty} |f(x)|^p dx \right)^{\frac{1}{p}} < \infty \right\},$$

$$g \in L^q(\mathbf{R}_+),$$

$\|f\|_p, \|g\|_q > 0$, the following Hardy–Hilbert integral inequality and its equivalent form hold true:

$$\int_0^{\infty} \int_0^{\infty} \frac{f(x)g(y)}{x+y} dx\, dy < \frac{\pi}{\sin(\pi/p)} \|f\|_p \|g\|_q, \tag{1.5}$$

$$\left[\int_0^{\infty} \left(\int_0^{\infty} \frac{f(x)}{x+y} dx \right)^p dy \right]^{\frac{1}{p}} < \frac{\pi}{\sin(\pi/p)} \|f\|_p, \tag{1.6}$$

where the constant factor

$$\frac{\pi}{\sin(\pi/p)}$$

is the best possible. If $a_m, b_n \geq 0$,

$$a = \{a_m\}_{m=1}^{\infty} \in l^p = \left\{ a; \|a\|_p = \left(\sum_{m=1}^{\infty} |a_m|^p \right)^{\frac{1}{p}} < \infty \right\},$$

$$b = \{b_n\}_{n=1}^{\infty} \in l^q,$$

$\|a\|_p, \|b\|_q > 0$, then we also have the following equivalent discrete variant of the above inequalities:

$$\sum_{n=1}^{\infty} \sum_{m=1}^{\infty} \frac{a_m b_n}{m+n} < \frac{\pi}{\sin(\pi/p)} \|a\|_p \|b\|_q, \tag{1.7}$$

$$\left[\sum_{n=1}^{\infty} \left(\sum_{m=1}^{\infty} \frac{a_m}{m+n} \right)^p \right]^{\frac{1}{p}} < \frac{\pi}{\sin(\pi/p)} \|a\|_p, \tag{1.8}$$

with the same best possible constant factor

$$\frac{\pi}{\sin(\pi/p)}.$$

For $p = q = 2$, inequalities (1.5)–(1.8) reduce, respectively, to (1.1)–(1.4).

In 1934, Hardy *et al.* [6] established some extensions of (1.5)–(1.8) as follows. Suppose that $p > 1, \frac{1}{p} + \frac{1}{q} = 1, k_1(x, y)$ is a nonnegative homogeneous function of degree -1, satisfying $k_1(ux, uy) = u^{-1} k_1(x, y)(u, x, y > 0)$, and

$$k_p := \int_0^{\infty} k_1(u, 1) u^{\frac{-1}{p}} \, du \in \mathbf{R}_+.$$

If

$$f(x), \ g(y) \geq 0, \ f \in L^p(\mathbf{R}_+), \ g \in L^q(\mathbf{R}_+), \ \|f\|_p, \ \|g\|_q > 0,$$

then we have the following Hardy–Hilbert-type integral inequality and its equivalent form:

$$\int_0^{\infty} \int_0^{\infty} k_1(x, y) f(x) g(y) dx \, dy < k_p \|f\|_p \|g\|_q, \tag{1.9}$$

$$\left[\int_0^{\infty} \left(\int_0^{\infty} k_1(x, y) f(x) dx \right)^p dy \right]^{\frac{1}{p}} < k_p \|f\|_p, \tag{1.10}$$

where the constant factor k_p is the best possible.

If $k_1(m,n)m^{-\frac{1}{p}}$ (resp. $k_1(m,n)n^{-\frac{1}{q}}$) is decreasing with respect to m (resp. n) $\in \mathbf{N}$, $a_m, b_n \geq 0$, $a = \{a_m\}_{m=1}^{\infty} \in l^p$, $b = \{b_n\}_{n=1}^{\infty} \in l^q$, $||a||_p, ||b||_q > 0$, then we still have an equivalent discrete variant of the above inequalities as follows:

$$\sum_{n=1}^{\infty}\sum_{m=1}^{\infty} k_1(m,n)a_m b_n < k_p ||a||_p ||b||_q, \qquad (1.11)$$

$$\left[\sum_{n=1}^{\infty}\left(\sum_{m=1}^{\infty} k_1(m,n)a_m\right)^p\right]^{\frac{1}{p}} < k_p ||a||_p, \qquad (1.12)$$

with the same best possible constant factor k_p.

For

$$k_1(x,y) = \frac{1}{x+y} \quad \left(\text{resp. } k_1(m,n) = \frac{1}{m+n}\right),$$

(1.9) and (1.10) (resp. (1.11) and (1.12)) reduce to (1.5) and (1.6) (resp. (1.7) and (1.8)), respectively.

Some applications of the above Hardy–Hilbert-type inequalities are featured in Ref. [7].

1.2 Hilbert-Type Inequalities with Multi-Parameters

In 1998, by introducing an independent parameter $\lambda \in (0,\infty)$, Yang ([8,9]) presented the following extension of (1.1).

If $f(x) \geq 0$ satisfying

$$0 < \int_0^{\infty} x^{1-\lambda} f^2(x) dx < \infty$$

and $g(y) \geq 0$ satisfying

$$0 < \int_0^{\infty} y^{1-\lambda} g^2(y) dy < \infty,$$

then we have

$$\int_0^\infty \int_0^\infty \frac{f(x)g(y)}{(x+y)^\lambda} dx\, dy < k_\lambda$$

$$\times \left(\int_0^\infty x^{1-\lambda} f^2(x) dx \int_0^\infty y^{1-\lambda} g^2(y) dy \right)^{\frac{1}{2}}, \qquad (1.13)$$

where the constant factor

$$k_\lambda = B\left(\frac{\lambda}{2}, \frac{\lambda}{2} \right)$$

is the best possible (where

$$B(u,v) := \int_0^\infty \frac{t^{u-1}}{(1+t)^{u+v}} dt \quad (u, v > 0)$$

is the beta function).

For $\lambda = 1$, inequality (1.13) reduces to (1.1).

In 2004, by introducing two pairs of conjugate exponents (p, q), (r, s) $(\frac{1}{p} + \frac{1}{q} = \frac{1}{r} + \frac{1}{s} = 1)$, Yang [10] proved the following extension of (1.5).

For $p, r > 1$, $f(x) \geq 0$ satisfying

$$0 < \int_0^\infty x^{p(1-\frac{\lambda}{r})-1} f^p(x) dx < \infty$$

and $g(y) \geq 0$ satisfying

$$0 < \int_0^\infty y^{q(1-\frac{\lambda}{s})-1} g^q(y) dy < \infty,$$

we have

$$\int_0^\infty \int_0^\infty \frac{f(x)g(y)}{x^\lambda + y^\lambda} dx\, dy$$

$$< \frac{\pi}{\lambda \sin(\pi/r)} \left[\int_0^\infty x^{p(1-\frac{\lambda}{r})-1} f^p(x) dx \right]^{\frac{1}{p}}$$

$$\times \left[\int_0^\infty y^{q(1-\frac{\lambda}{s})-1} g^q(y) dy \right]^{\frac{1}{q}}, \qquad (1.14)$$

where the constant factor

$$\frac{\pi}{\lambda \sin(\pi/r)}$$

is the best possible.

For $\lambda = 1, r = p$, and $s = q$, (1.14) reduces to the dual form of (1.5) as follows:

$$\int_0^\infty \int_0^\infty \frac{f(x)g(y)}{x+y} dx\, dy$$

$$< \frac{\pi}{\sin(\pi/p)} \left(\int_0^\infty x^{p-2} f^p(x) dx \right)^{\frac{1}{p}} \left(\int_0^\infty y^{q-2} g^q(y) dy \right)^{\frac{1}{q}}. \quad (1.15)$$

For $p = q = 2$, both (1.5) and (1.15) reduce to (1.1).

Recently, by introducing multi-parameters $\lambda_1, \lambda_2, \lambda \in \mathbf{R}$ $(\lambda_1 + \lambda_2 = \lambda)$, Yang [11,12] established the following extensions of (1.9)–(1.13).

Suppose that $k_\lambda(x,y)$ is a nonnegative homogeneous function of degree $-\lambda$ in $\mathbf{R}_+^2$, satisfying

$$k_\lambda(ux, uy) = u^{-\lambda} k_\lambda(x,y)(u,x,y>0)$$

and

$$k(\lambda_1) := \int_0^\infty k_\lambda(t,1) t^{\lambda_1-1} dt \in \mathbf{R}_+,$$

$$\phi(x) = x^{p(1-\lambda_1)-1}, \quad \psi(y) = y^{q(1-\lambda_2)-1}(x,y \in \mathbf{R}_+).$$

If $f(x), g(y) \geq 0$,

$$f \in L_{p,\phi}(\mathbf{R}_+) = \left\{ f; \|f\|_{p,\phi} := \left(\int_0^\infty \phi(x) |f(x)|^p dx \right)^{\frac{1}{p}} < \infty \right\},$$

$$g \in L_{q,\psi}(\mathbf{R}_+),$$

$\|f\|_{p,\phi}, \|g\|_{q,\psi} > 0$, then we have the following Hilbert-type integral inequality and its equivalent form with multi-parameters:

$$\int_0^\infty \int_0^\infty k_\lambda(x,y) f(x) g(y) dx\, dy < k(\lambda_1) \|f\|_{p,\phi} \|g\|_{q,\psi}, \quad (1.16)$$

$$\left[\int_0^\infty \left(\int_0^\infty k_\lambda(x,y) f(x) dx \right)^p dy \right]^{\frac{1}{p}} < k(\lambda_1) \|f\|_{p,\phi}, \quad (1.17)$$

where the constant factor $k(\lambda_1)$ is the best possible. Moreover, if $k_\lambda(m,n) m^{\lambda_1-1}$ (resp. $k_\lambda(m,n) n^{\lambda_2-1}$) is decreasing with respect to m

(resp. n), then for $a_m, b_n \geq 0$,

$$a \in l_{p,\phi} = \left\{ a; \|a\|_{p,\phi} := \left(\sum_{m=1}^{\infty} \phi(m)|a_m|^p \right)^{\frac{1}{p}} < \infty \right\},$$

$$b = \{b_n\}_{n=1}^{\infty} \in l_{q,\psi},$$

$\|a\|_{p,\phi}, \|b\|_{q,\psi} > 0$, we have the following equivalent discrete variant of the above inequalities:

$$\sum_{n=1}^{\infty} \sum_{m=1}^{\infty} k_\lambda(m,n) a_m b_n < k(\lambda_1) \|a\|_{p,\phi} \|b\|_{q,\psi}, \tag{1.18}$$

$$\left[\sum_{n=1}^{\infty} \left(\sum_{m=1}^{\infty} k_\lambda(m,n) a_m \right)^p \right]^{\frac{1}{p}} < k(\lambda_1) \|a\|_{p,\phi}, \tag{1.19}$$

where the constant factor $k(\lambda_1)$ is still the best possible.

Clearly, for $\lambda = 1$, $\lambda_1 = \frac{1}{q}$, and $\lambda_2 = \frac{1}{p}$, (1.16)–(1.19) reduce, respectively, to (1.9)–(1.12); for $p = q = 2$, $\lambda_1 = \lambda_2 = \frac{\lambda}{2} > 0$, and $k_\lambda(x,y) = \frac{1}{(x+y)^\lambda}$, (1.1) reduces to (1.13); for $\lambda_1 = \frac{\lambda}{r}, \lambda_2 = \frac{\lambda}{s}$, and $k_\lambda(x,y) = \frac{1}{x^\lambda + y^\lambda}$ ($\lambda > 0$), (1.16) reduces to (1.14).

On half-discrete Hilbert-type inequalities with nonhomogeneous kernels, Hardy *et al.* proved a few results in Theorem 351 of Ref. [6]. But they did not prove that the constant factors are the best possible. However, Yang [13] presented a result with the kernel

$$\frac{1}{(1 + nx)^\lambda}$$

by introducing a variable and proved that the corresponding constant factor is the best possible.

Using weight functions, the following half-discrete Hilbert-type inequality and its equivalent forms with a general homogeneous kernel of degree $-\lambda \in \mathbf{R}$ and a best possible constant factor $k(\lambda_1)$ is obtained (cf. Ref. [14]):

$$\int_0^\infty f(x) \sum_{n=1}^\infty k_\lambda(x,n) a_n dx < k(\lambda_1) ||f||_{p,\phi} ||a||_{q,\psi}, \qquad (1.20)$$

$$\left[\sum_{n=1}^\infty n^{p\lambda_2 - 1} \left(\int_0^\infty k_\lambda(x,n) f(x) dx \right)^p \right]^{\frac{1}{p}} < k(\lambda_1) ||f||_{p,\phi}, \qquad (1.21)$$

$$\left[\int_0^\infty x^{q\lambda_1 - 1} \left(\sum_{n=1}^\infty k_\lambda(x,n) a_n \right)^q dx \right]^{\frac{1}{q}} < k(\lambda_1) ||a||_{q,\psi}. \qquad (1.22)$$

A half-discrete Hilbert-type inequality with a general nonhomogeneous kernel $k_\lambda(1, xn)$ and a best possible constant factor were obtained by Yang *et al.* [15–17]. Surveys on Hilbert-type inequalities with multi-parameters are presented in Refs. [18,19].

1.3 Multidimensional Hilbert-Type Inequalities and a Remark

If $i_0, j_0 \in \mathbf{N}$, $\alpha, \beta > 0$, we set

$$||x||_\alpha := \left(\sum_{k=1}^{i_0} |x_k|^\alpha \right)^{\frac{1}{\alpha}} \quad (x = (x_1, \ldots, x_{i_0}) \in \mathbf{R}^{i_0}),$$

$$||y||_\beta := \left(\sum_{k=1}^{j_0} |y_k|^\beta \right)^{\frac{1}{\beta}} \quad (y = (y_1, \ldots, y_{j_0}) \in \mathbf{R}^{j_0}).$$

In 2006, by using the transfer formula, Hong [20] presented a multidimensional Hilbert-type integral inequality and its equivalent form (for $\beta = \alpha$) as follows.

For $p > 1$, $\frac{1}{p} + \frac{1}{q} = 1$, $\lambda_1 + \lambda_2 = \lambda$,

$$\Phi(x) = x^{p(i_0 - \lambda_1) - i_0} (x \in \mathbf{R}_+^{i_0}), \quad \Psi(y) = y^{p(j_0 - \lambda_2) - j_0} (y \in \mathbf{R}_+^{j_0}),$$

$f(x) = f(x_1, \ldots, x_{i_0}) \geq 0$, $g(y) = g(y_1, \ldots, y_{j_0}) \geq 0$,

$$0 < ||f||_{p,\Phi} = \left(\int_{\mathbf{R}_+^{i_0}} \Phi(x) f^p(x) dx \right)^{\frac{1}{p}} < \infty,$$

$$0 < ||g||_{q,\Psi} = \left(\int_{\mathbf{R}_+^{j_0}} \Psi(y) g^q(y) dy \right)^{\frac{1}{q}} < \infty,$$

we have the following equivalent inequalities with the kernel

$$k_\lambda(||x||_\alpha, ||y||_\alpha) = \frac{1}{(||x||_\alpha + ||y||_\alpha)^\lambda} \quad (\lambda > 0):$$

$$\int_{\mathbf{R}_+^{j_0}} \int_{\mathbf{R}_+^{i_0}} k_\lambda(||x||_\alpha, ||y||_\beta) f(x) g(y) dx\, dy < K(\lambda_1) ||f||_{p,\Phi} ||g||_{q,\Psi},$$

(1.23)

$$\left[\int_{\mathbf{R}_+^{j_0}} ||y||_\beta^{\frac{p\lambda}{s}-1} \left(\int_{\mathbf{R}_+^{i_0}} k_\lambda(||x||_\alpha, ||y||_\beta) f(x) dx \right) dy \right]^{\frac{1}{p}} < K(\lambda_1) ||f||_{p,\Phi},$$

(1.24)

where the constant factor

$$K(\lambda_1) = \frac{\Gamma^{i_0}(1/\alpha)}{\alpha^{i_0-1}\Gamma(i_0/\alpha)} B(\lambda_1, \lambda_2)$$

$(\lambda_1, \lambda_2 > 0)$ is the best possible.

In 2007, by introducing four particular kernels

$$k_\lambda(||x||_\alpha, ||y||_\alpha) = \frac{1}{|||x||_\alpha - ||y||_\alpha|^\lambda} \quad (0 < \lambda < 1),$$

$$k_\lambda(||x||_\alpha, ||y||_\alpha) = \frac{\ln(||x||_\alpha/||y||_\alpha)}{||x||_\alpha^\lambda - ||y||_\alpha^\lambda} \quad (\lambda > 0),$$

$$k_\lambda(||x||_\alpha, ||y||_\alpha) = \frac{1}{||x||_\alpha^\lambda + ||y||_\alpha^\lambda} \quad (\lambda > 0),$$

and

$$k_\lambda(||x||_\alpha, ||y||_\alpha) = \frac{1}{(\max\{||x||_\alpha, ||y||_\alpha\})^\lambda} \quad (\lambda > 0),$$

Zhong and Yang [39] proved the four pairs of equivalent inequalities (1.23) and (1.24) (for $\beta = \alpha$) with the particular best possible constant factors

$$K(\lambda_1) = \frac{\Gamma^{i_0}(1/\alpha)}{\alpha^{i_0-1}\Gamma(i_0/\alpha)} k(\lambda_1),$$

where

$$k(\lambda_1) = B(\lambda_1, 1-\lambda) + B(\lambda_2, 1-\lambda), \left[\frac{\pi}{\lambda \sin(\pi\lambda_1/\lambda)}\right]^2, \frac{\pi}{\lambda \sin(\pi\lambda_1/\lambda)},$$

and $\frac{\lambda}{\lambda_1\lambda_2}$ $(\lambda_1, \lambda_2 > 0)$.

In 2011–2012, Yang and Krnić [22] proved (1.23) and (1.24) with the general homogeneous kernel $k_\lambda(||x||_\alpha, ||y||_\beta)$ and the best possible constant factor

$$K(\lambda_1) = \left(\frac{\Gamma^{j_0}(1/\beta)}{\beta^{j_0-1}\Gamma(j_0/\beta)}\right)^{\frac{1}{p}} \left(\frac{\Gamma^{i_0}(1/\alpha)}{\alpha^{i_0-1}\Gamma(i_0/\alpha)}\right)^{\frac{1}{q}} k(\lambda_1),$$

where

$$k(\lambda_1) = \int_0^\infty k_\lambda(t, 1) t^{\lambda_1-1} dt \in \mathbf{R}_+.$$

In this case, for $i_0 = j_0 = \alpha = \beta_1$, (1.23) and (1.24) reduce, respectively, to (1.16) and (1.17).

In recent years, some results on multidimensional Hilbert-type integral inequalities have been published in Refs. [23–27,39] and some results on multidimensional discrete, half-discrete Hilbert-type inequalities are provided in Refs. [28–49]. A book about the study of this type of inequalities was published in 2020 by Yang *et al.* [48].

Remark 1.1. (1) Many different kinds of Hilbert-type discrete, half-discrete, and integral inequalities with applications have been presented during the past 20 years. In the current book, special attention is given to new results proved during 2009–2020. Included within are many generalizations, extensions, and refinements of Hilbert-type discrete, half-discrete, and integral inequalities involving many special functions, such as the beta function, the gamma function, hypergeometric functions, trigonometric functions, hyperbolic functions, the Hurwitz zeta function, the Riemann zeta function, Bernoulli functions, as well as Bernoulli numbers and Euler's constant (cf. Refs. [49–83]).

(2) In his nine books [11,12,84–90], Yang presented several kinds of Hilbert-type operators with general homogeneous and nonhomogeneous kernels and two pairs of conjugate exponents as well as related inequalities. These research monographs contained recent developments of discrete, half-discrete, and integral types of operators and inequalities with proofs, examples, and applications.[1]

(3) In 2017, Hong [91] studied an equivalent statement between a Hilbert-type integral inequality with a general homogeneous kernel and some parameters. Some authors continue to study this topic for other types of integral inequalities and operators (cf. Refs. [92–107]). This idea helps us to consider the equivalent properties in several kinds of Hilbert-type and Hardy-type integral inequalities and applications to the extended Hurwitz zeta function and the Riemann zeta function.

(4) In 2006, using the Euler–Maclaurin summation formula, Krnić *et al.* [108] presented an extension of (1.7) with the kernel

$$\frac{1}{(m+n)^{\lambda}} \quad (0 < \lambda \le 4).$$

In 2019–2020, applying the results reported in Ref. [108], Adiyasuren *et al.* [109] considered a Hilbert-type inequality involving partial sums, and subsequently, Mo *et al.* [110] presented a Hilbert-type integral inequality involving two upper-limit functions. Some additional results are available in Refs. [111–118].

In the following Chapter 2, by using weight functions and techniques of real and functional analyses, we consider the equivalent properties of a new Hilbert-type integral inequality with a nonhomogeneous kernel and the case of homogeneous kernel, operator expressions, reverses, and applications related to the Riemann zeta function and the Hurwitz zeta function. In Chapters 3–5, applying the extended Hardy–Hilbert inequalities and the gamma and beta functions, we obtain some new Hilbert-type integral inequalities involving the derivative functions and the upper-limit functions. In Chapter 6, we return to the study of the equivalent properties of the cases in Hardy-type integral inequalities and their applications to the extended Hurwitz-zeta function.

---

[1]On a different note, for the study of other types of interesting operators, the interested reader is also referred to Ref. [126].

Chapter 2

Equivalent Properties of a New Hilbert-Type Integral Inequality with Parameters

In this chapter, by introducing independent parameters using methods of real analysis and weight functions, we prove a new Hilbert-type integral inequality with a general nonhomogeneous kernel. Equivalent statements related to the best possible constant factor and several parameters are considered. We also deduce the cases of a homogeneous kernel and some particular examples. In the form of applications, we consider the operator expressions, some particular cases involving the Riemann zeta function and the Hurwitz zeta function, as well as reverses.

2.1 Some Lemmas

Hereinafter in this chapter, we suppose that $h(u)$ is a nonnegative measurable function in $\mathbf{R}_+ = (0, \infty)$, $p > 0$ $(p \neq 1), \frac{1}{p} + \frac{1}{q} = 1$, $\gamma = \sigma, \sigma_1 \in \mathbf{R} = (-\infty, \infty)$ such that

$$k(\gamma) := \int_0^\infty h(u)u^{\gamma-1}du \in \mathbf{R}_+.$$

We also assume that $f(x)$ and $g(y)$ are nonnegative measurable functions in $\mathbf{R}_+$, satisfying

$$0 < \int_0^\infty x^{p[1-(\frac{\sigma}{p}+\frac{\sigma_1}{q})]-1} f^p(x)dx < \infty \quad \text{and}$$

$$0 < \int_0^\infty y^{q[1-(\frac{\sigma}{p}+\frac{\sigma_1}{q})]-1} g^q(y)dy < \infty. \tag{2.1}$$

Lemma 2.1. *If there exists a constant $\delta_0 > 0$ such that $k(\sigma \pm \delta_0) < \infty$, then the function $k(\eta)$ is continuous in any $\eta \in (\sigma - \delta_0, \sigma + \delta_0)$, satisfying*

$$0 \le k(\eta) \le k(\sigma - \delta_0) + k(\sigma + \delta_0).$$

Proof.   For any sequence

$$\{\delta_n\}_{n=1}^\infty \subset [\sigma - \delta_0 - \eta, \sigma + \delta_0 - \eta], \delta_n \to 0 \ (n \to \infty),$$

we have

$$\sigma - \delta_0 \le \eta + \delta_n \le \sigma + \delta_0 \ (n \in \mathbf{N}),$$

$$k(\eta + \delta_n) = \int_0^1 h(u)u^{\eta+\delta_n-1}du + \int_1^\infty h(u)u^{\eta+\delta_n-1}du$$

$$\le \int_0^1 h(u)u^{\sigma-\delta_0-1}du + \int_1^\infty h(u)u^{\sigma+\delta_0-1}du$$

$$\le \int_0^\infty h(u)u^{\sigma-\delta_0-1}du + \int_0^\infty h(u)u^{\sigma+\delta_0-1}du$$

$$= k(\sigma - \delta_0) + k(\sigma + \delta_0).$$

We indicate the dominated function $F(u)$ as follows:

$$F(u) := \begin{cases} h(u)u^{\sigma-\delta_0-1}, & u \in (0,1], \\ h(u)u^{\sigma+\delta_0-1}, & u \in (1,\infty), \end{cases}$$

thus

$$|f_n(u)| = f_n(u) := h(u)u^{\eta+\delta_n-1} \le F(u)(u \in (0,\infty)),$$

$$0 < \int_0^\infty F(u)du \le k(\sigma - \delta_0) + k(\sigma + \delta_0) < \infty.$$

By Lebesgue's dominated convergence theorem (cf. Ref. [119]), we have

$$k(\eta + \delta_n) = \int_0^\infty f_n(u)du = \int_0^\infty h(u)u^{\eta+\delta_n-1}du$$

$$\to \int_0^\infty h(u)u^{\eta-1}du = k(\eta) \ (n \to \infty).$$

Hence, the function $k(\eta)$ is continuous in $\eta \in (\sigma-\delta_0, \sigma+\delta_0)$, satisfying

$$0 \le k(\eta) \le \int_0^\infty F(u)du \le k(\sigma - \delta_0) + k(\sigma + \delta_0).$$

This completes the proof of the lemma.   □

Lemma 2.2. *For $p > 1$ $(q > 1)$, we have the following Hilbert-type integral inequality with a general nonhomogeneous kernel:*

$$I := \int_0^\infty \int_0^\infty h(xy)f(x)g(y)dx\,dy < k^{\frac{1}{p}}(\sigma)k^{\frac{1}{q}}(\sigma_1)$$

$$\times \left\{ \int_0^\infty x^{p[1-(\frac{\sigma}{p}+\frac{\sigma_1}{q})]-1}f^p(x)dx \right\}^{\frac{1}{p}}$$

$$\times \left\{ \int_0^\infty y^{q[1-(\frac{\sigma}{p}+\frac{\sigma_1}{q})]-1}g^q(y)dy \right\}^{\frac{1}{q}}. \tag{2.2}$$

Proof. We define the following weight function:

$$\omega(\sigma, x) := x^\sigma \int_0^\infty h(xy)y^{\sigma-1}dy \ (x \in \mathbf{R}_+). \tag{2.3}$$

Setting $u = xy$, we obtain

$$\omega(\sigma, x) = \int_0^\infty h(u)u^{\sigma-1}du = k(\sigma). \tag{2.4}$$

Similarly, it follows that

$$\omega(\sigma_1, y) = y^{\sigma_1} \int_0^\infty h(xy)x^{\sigma_1-1}dx = k(\sigma_1). \tag{2.5}$$

By Hölder's inequality (cf. Ref. [120]), (2.3) and (2.5), we have

$$I = \int_0^\infty \int_0^\infty h(xy) \left[\frac{y^{(\sigma-1)/p}}{x^{(\sigma_1-1)/q}} f(x) \right] \left[\frac{x^{(\sigma_1-1)/q}}{y^{(\sigma-1)/p}} g(y) \right] dx\,dy$$

$$\leq \left\{ \int_0^\infty \int_0^\infty h(xy) \frac{y^{\sigma-1}}{x^{(\sigma_1-1)(p-1)}} f^p(x) dx\,dy \right\}^{\frac{1}{p}}$$

$$\times \left\{ \int_0^\infty \int_0^\infty h(xy) \frac{x^{\sigma_1-1}}{y^{(\sigma-1)(q-1)}} g^q(y) dx\,dy \right\}^{\frac{1}{q}}$$

$$= \left\{ \int_0^\infty \omega(\sigma, x) x^{p[1-(\frac{\sigma}{p}+\frac{\sigma_1}{q})]-1} f^p(x) dx \right\}^{\frac{1}{p}}$$

$$\times \left\{ \int_0^\infty \omega(\sigma_1, y) y^{q[1-(\frac{\sigma}{p}+\frac{\sigma_1}{q})]-1} g^q(y) dy \right\}^{\frac{1}{q}}. \tag{2.6}$$

If (2.6) keeps the form of equality, then there exist constants A and B such that they are not both zero and (cf. Ref. [120])

$$A \frac{y^{\sigma-1}}{x^{(\sigma_1-1)(p-1)}} f^p(x) = B \frac{x^{\sigma_1-1}}{y^{(\sigma-1)(q-1)}} g^q(y) \quad \text{a.e. in } (0, \infty) \times (0, \infty).$$

Assuming that $A \neq 0$, there exists a $y \in (0, \infty)$ such that

$$x^{p[1-(\frac{\sigma}{p}+\frac{\sigma_1}{q})]-1} f^p(x) = \frac{Bg^q(y)}{Ay^{(\sigma-1)q}} x^{\sigma_1-\sigma-1} \quad \text{a.e. in } (0, \infty).$$

Since

$$\int_0^\infty x^{\sigma_1-\sigma-1} dx = \infty,$$

the above expression contradicts the fact that

$$0 < \int_0^\infty x^{p[1-(\frac{\sigma}{p}+\frac{\sigma_1}{q})]-1} f^p(x) dx < \infty.$$

By (2.4) and (2.5), we derive (2.2).

This completes the proof of the lemma.   □

Remark 2.3. (i) If $\sigma_1 = \sigma$, then in view of (2.1) and (2.2), we have

$$0 < \int_0^\infty x^{p(1-\sigma)-1} f^p(x)dx < \infty, \quad 0 < \int_0^\infty y^{q(1-\sigma)-1} g^q(y)dy < \infty$$

and the following inequality:

$$I = \int_0^\infty \int_0^\infty h(xy)f(x)g(y)dx\,dy$$

$$< k(\sigma) \left[\int_0^\infty x^{p(1-\sigma)-1} f^p(x)dx\right]^{\frac{1}{p}} \left[\int_0^\infty y^{q(1-\sigma)-1} g^q(y)dy\right]^{\frac{1}{q}}.$$

$$(2.7)$$

(ii) For $0 < p < 1$ ($q < 0$), by the reverse Hölder inequality (cf. Ref. [120]) and in the same manner, we obtain the reverse of (2.2) and (2.7).

Lemma 2.4. *For $p > 1$, the constant factor in (2.7) is the best possible.*

Proof.   For any $\varepsilon > 0$, we set

$$\widetilde{f}(x) := \begin{cases} 0, & 0 < x < 1, \\ x^{\sigma - \frac{\varepsilon}{p} - 1}, & x \geq 1, \end{cases} \quad \text{and}$$

$$\widetilde{g}(y) := \begin{cases} y^{\sigma + \frac{\varepsilon}{q} - 1}, & 0 < y \leq 1, \\ 0, & y > 1. \end{cases}$$

If there exists a positive constant M ($\leq k(\sigma)$) such that (2.7) is valid when replacing $k(\sigma)$ by M, then in particular, we have

$$\widetilde{I} := \int_0^\infty \int_0^\infty h(xy)\widetilde{f}(x)\widetilde{g}(y)dx\,dy$$

$$< M \left[\int_0^\infty x^{p(1-\sigma)-1} \widetilde{f}^p(x)dx\right]^{\frac{1}{p}} \left[\int_0^\infty y^{q(1-\sigma)-1} \widetilde{g}^q(y)dy\right]^{\frac{1}{q}}$$

$$= M \left(\int_1^\infty x^{-\varepsilon-1}dx\right)^{\frac{1}{p}} \left(\int_0^1 y^{\varepsilon-1}dy\right)^{\frac{1}{q}} = \frac{1}{\varepsilon}.$$

By Fubini's theorem (cf. Ref. [119]), we have

$$\tilde{I} = \int_0^1 \left(\int_1^\infty h(xy) x^{\sigma - \frac{\varepsilon}{p} - 1} dx \right) y^{\sigma + \frac{\varepsilon}{q} - 1} dy$$

$$\overset{u=xy}{=} \int_0^1 \left(\int_y^\infty h(u) u^{\sigma - \frac{\varepsilon}{p} - 1} du \right) y^{\varepsilon - 1} dy$$

$$= \int_0^1 \left(\int_y^1 h(u) u^{\sigma - \frac{\varepsilon}{p} - 1} du \right) y^{\varepsilon - 1} dy$$

$$+ \int_0^1 \left(\int_1^\infty h(u) u^{\sigma - \frac{\varepsilon}{p} - 1} du \right) y^{\varepsilon - 1} dy$$

$$= \int_0^1 \left(\int_0^u y^{\varepsilon - 1} dy \right) h(u) u^{\sigma - \frac{\varepsilon}{p} - 1} du + \frac{1}{\varepsilon} \int_1^\infty h(u) u^{\sigma - \frac{\varepsilon}{p} - 1} du$$

$$= \frac{1}{\varepsilon} \left(\int_0^1 h(u) u^{\sigma + \frac{\varepsilon}{q} - 1} du + \int_1^\infty h(u) u^{\sigma - \frac{\varepsilon}{p} - 1} du \right).$$

In view of the above results, it follows that

$$\int_0^1 h(u) u^{\sigma + \frac{\varepsilon}{q} - 1} du + \int_1^\infty h(u) u^{\sigma - \frac{\varepsilon}{p} - 1} du = \varepsilon \tilde{I} < M.$$

For $\varepsilon \to 0^+$, by Fatou's lemma (cf. Ref. [119]), we derive that

$$k(\sigma) = \int_0^1 \lim_{\varepsilon \to 0^+} h(u) u^{\sigma + \frac{\varepsilon}{q} - 1} du + \int_1^\infty \lim_{\varepsilon \to 0^+} h(u) u^{\sigma - \frac{\varepsilon}{p} - 1} du$$

$$\leq \lim_{\varepsilon \to 0^+} \left[\int_0^1 h(u) u^{\sigma + \frac{\varepsilon}{q} - 1} du + \int_1^\infty h(u) u^{\sigma - \frac{\varepsilon}{p} - 1} du \right] \leq M.$$

Hence, $M = k(\sigma)$ is the best possible constant factor in (2.7). This completes the proof of the lemma. $\square$

Remark 2.5. We set a parameter

$$\hat{\sigma} := \frac{\sigma}{p} + \frac{\sigma_1}{q}.$$

We can rewrite (2.2) as follows:

$$\int_0^\infty \int_0^\infty h(xy)f(x)g(y)dx\,dy$$

$$< k^{\frac{1}{p}}(\sigma)k^{\frac{1}{q}}(\sigma_1)\left[\int_0^\infty x^{p(1-\widehat{\sigma})-1}f^p(x)dx\right]^{\frac{1}{p}}$$

$$\times \left[\int_0^\infty y^{q(1-\widehat{\sigma})-1}g^q(y)dy\right]^{\frac{1}{q}}. \tag{2.8}$$

By Hölder's inequality with weight (cf. Ref. [119]), we obtain the following inequality:

$$0 < k(\widehat{\sigma}) = k\left(\frac{\sigma}{p} + \frac{\sigma_1}{q}\right) = \int_0^\infty h(u)u^{\frac{\sigma}{p}+\frac{\sigma_1}{q}-1}du$$

$$= \int_0^\infty h(u)(u^{\frac{\sigma-1}{p}})(u^{\frac{\sigma_1-1}{q}})du$$

$$\leq \left(\int_0^\infty h(u)u^{\sigma-1}du\right)^{\frac{1}{p}}\left(\int_0^\infty h(u)u^{\sigma_1-1}du\right)^{\frac{1}{q}}$$

$$= k^{\frac{1}{p}}(\sigma)k^{\frac{1}{q}}(\sigma_1) < \infty. \tag{2.9}$$

Lemma 2.6. *For $p > 1$, if the constant factor $k^{\frac{1}{p}}(\sigma)k^{\frac{1}{q}}(\sigma_1)$ in (2.8) (or (2.2)) is the best possible, then $\sigma_1 = \sigma$.*

Proof. If the constant factor $k^{\frac{1}{p}}(\sigma)k^{\frac{1}{q}}(\sigma_1)$ in (2.8) is the best possible, then in view of (2.7) (for $\sigma = \widehat{\sigma}$), we have

$$k^{\frac{1}{p}}(\sigma)k^{\frac{1}{q}}(\sigma_1) \leq k(\widehat{\sigma}) \ (\in \mathbf{R}_+),$$

namely, (2.9) keeps the form of equality.

We observe that (2.9) keeps the form of equality if and only if there exist constants A and B such that they are not both zero and (cf. Ref. [120])

$$Au^{\sigma-1} = Bu^{\sigma_1-1} \quad \text{a.e. in } \mathbf{R}_+.$$

Assuming that $A \neq 0$, we have

$$u^{\sigma - \sigma_1} = \frac{B}{A} \quad \text{a.e. in } \mathbf{R}_+,$$

thus

$$\sigma - \sigma_1 = 0, \quad \text{namely, } \sigma_1 = \sigma.$$

This completes the proof of the lemma. $\qquad\square$

2.2   Main Results and Some Corollaries

Theorem 2.7. *For $p > 1$, inequality (2.2) is equivalent to the following inequality:*

$$J := \left[\int_0^\infty y^{p(\frac{\sigma}{p} + \frac{\sigma_1}{q}) - 1} \left(\int_0^\infty h(xy) f(x) dx \right)^p dy \right]^{\frac{1}{p}}$$

$$< k^{\frac{1}{p}}(\sigma) k^{\frac{1}{q}}(\sigma_1) \left\{ \int_0^\infty x^{p[1 - (\frac{\sigma}{p} + \frac{\sigma_1}{q})] - 1} f^p(x) dx \right\}^{\frac{1}{p}}. \quad (2.10)$$

The constant factor

$$k^{\frac{1}{p}}(\sigma) k^{\frac{1}{q}}(\sigma_1)$$

in (2.10) is the best possible if and only if the same constant factor in (2.2) is the best possible.

In particular, for $\sigma_1 = \sigma$, we have the following inequality equivalent to (2.7) with the same best possible constant factor $k(\sigma)$:

$$\left[\int_0^\infty y^{p\sigma - 1} \left(\int_0^\infty h(xy) f(x) dx \right)^p dy \right]^{\frac{1}{p}}$$

$$< k(\sigma) \left[\int_0^\infty x^{p(1 - \sigma) - 1} f^p(x) dx \right]^{\frac{1}{p}}. \quad (2.11)$$

Proof.   If (2.10) is valid, then by Hölder's inequality, we have

$$I = \int_0^\infty \left[y^{\frac{-1}{p} + (\frac{\sigma}{p} + \frac{\sigma_1}{q})} \int_0^\infty h(xy) f(x) dx \right] \left[y^{\frac{1}{p} - (\frac{\sigma}{p} + \frac{\sigma_1}{q})} g(y) \right] dy$$

$$\leq J \left\{ \int_0^\infty y^{q[1 - (\frac{\sigma}{p} + \frac{\sigma_1}{q})] - 1} g^q(y) dy \right\}^{\frac{1}{q}}. \quad (2.12)$$

By (2.10), we derive (2.2).

On the other hand, assuming that (2.2) is valid, we set

$$g(y) := y^{p(\frac{\sigma}{p}+\frac{\sigma_1}{q})-1}\left(\int_0^\infty h(xy)f(x)dx\right)^{p-1} \quad (y > 0).$$

Then, it follows that

$$J^p = \int_0^\infty y^{q[1-(\frac{\sigma}{p}+\frac{\sigma_1}{q})]-1}g^q(y)dy = I. \tag{2.13}$$

If $J = 0$, then (2.10) is naturally valid; if $J = \infty$, then it is impossible to make (2.10) valid, namely, $J < \infty$. Suppose that $0 < J < \infty$. By (2.2), we have

$$J^p = \int_0^\infty y^{q[1-(\frac{\sigma}{p}+\frac{\sigma_1}{q})]-1}g^q(y)dy = I$$

$$< k^{\frac{1}{p}}(\sigma)k^{\frac{1}{q}}(\sigma_1)\left\{\int_0^\infty x^{p[1-(\frac{\sigma}{p}+\frac{\sigma_1}{q})]-1}f^p(x)dx\right\}^{\frac{1}{p}}J^{p-1},$$

$$J = \left\{\int_0^\infty y^{q[1-(\frac{\sigma}{p}+\frac{\sigma_1}{q})]-1}g^q(y)dy\right\}^{\frac{1}{p}}$$

$$< k^{\frac{1}{p}}(\sigma)k^{\frac{1}{q}}(\sigma_1)\left\{\int_0^\infty x^{p[1-(\frac{\sigma}{p}+\frac{\sigma_1}{q})]-1}f^p(x)dx\right\}^{\frac{1}{p}},$$

namely, (2.10) follows, which is equivalent to (2.2).

If the constant factor in (2.2) is the best possible, then the same constant factor in (2.10) is also the best possible. Otherwise, by (2.12), we would reach a contradiction that the same constant factor in (2.2) is not the best possible. Similarly, if the constant factor in (2.10) is the best possible, then the same constant factor in (2.2) is also the best possible. Otherwise, by (2.13), we would reach a contradiction that the same constant factor in (2.10) is not the best possible.

This completes the proof of the theorem. □

Theorem 2.8. *For $p > 1$, the following statements, (i), (ii), (iii), and (iv), are equivalent:*

(i) both

$$k^{\frac{1}{p}}(\sigma)k^{\frac{1}{q}}(\sigma_1) \quad \text{and} \quad k\left(\frac{\sigma}{p} + \frac{\sigma_1}{q}\right)$$

are independent of p and q;

(ii) $k^{\frac{1}{p}}(\sigma)k^{\frac{1}{q}}(\sigma_1) \leq k(\frac{\sigma}{p} + \frac{\sigma_1}{q})$;

(iii) $\sigma_1 = \sigma$;

(iv) the constant factor $k^{\frac{1}{p}}(\sigma)k^{\frac{1}{q}}(\sigma_1)$ in (2.2) and (2.10) is the best possible.

Proof.   (i) $\Rightarrow$ (ii): We have

$$k^{\frac{1}{p}}(\sigma)k^{\frac{1}{q}}(\sigma_1) = \lim_{p\to 1^+}\lim_{q\to\infty} k^{\frac{1}{p}}(\sigma)k^{\frac{1}{q}}(\sigma_1) = k(\sigma).$$

By Fatou's lemma (cf. Ref. [120]), we obtain that

$$k\left(\frac{\sigma}{p} + \frac{\sigma_1}{q}\right) = \lim_{p\to 1^+}\lim_{q\to\infty} k\left(\frac{\sigma}{p} + \frac{\sigma_1}{q}\right)$$

$$\geq k(\sigma) = k^{\frac{1}{p}}(\sigma)k^{\frac{1}{q}}(\sigma_1).$$

(ii) $\Rightarrow$ (iii): If

$$k^{\frac{1}{p}}(\sigma)k^{\frac{1}{q}}(\sigma_1) \leq k\left(\frac{\sigma}{p} + \frac{\sigma_1}{q}\right),$$

then (2.9) keeps the form of equality. In view of the proof of Lemma 2.6, we have $\sigma_1 = \sigma$.

(iii) $\Rightarrow$ (iv): By Lemma 2.4 and Theorem 2.7, the constant factor

$$k^{\frac{1}{p}}(\sigma)k^{\frac{1}{q}}(\sigma_1)(= k(\sigma))$$

is the best possible in (2.2) and (2.10).

(iv) $\Rightarrow$ (i): By Lemma 2.6, we have $\sigma_1 = \sigma$, and then, both $k^{\frac{1}{p}}(\sigma)k^{\frac{1}{q}}(\sigma_1)$ and $k(\frac{\sigma}{p} + \frac{\sigma_1}{q})$ are equal to $k(\sigma)$, which is independent of p and q.

Hence, statements (i), (ii), (iii), and (iv) are equivalent. This completes the proof of the theorem. $\square$

If $k_\lambda(x,y)(\geq 0)$ is a homogeneous function of degree $-\lambda$ satisfying

$$k_\lambda(ux, uy) = u^{-\lambda} k_\lambda(x,y) \quad (u, x, y > 0),$$

then setting

$$h(u) = k_\lambda(1, u)$$

and replacing x by $\frac{1}{x}$ and

$$x^{\lambda-2} f\left(\frac{1}{x}\right)$$

by $f(x)$ in Lemma 2.2, Theorem 2.7, and Theorem 2.8, for $\sigma_1 = \lambda - \mu$, we deduce the following corollary.

Corollary 2.9. *For $p > 1$, if*

$$k_\lambda(\gamma) := \int_0^\infty k_\lambda(1, u) u^{\gamma-1} du \in \mathbf{R}_+ (\gamma = \sigma, \lambda - \mu),$$

$$0 < \int_0^\infty x^{p[1-(\frac{\lambda-\sigma}{p}+\frac{\mu}{q})]-1} f^p(x) dx < \infty, \text{ and}$$

$$0 < \int_0^\infty y^{q[1-(\frac{\sigma}{p}+\frac{\lambda-\mu}{q})]-1} g^q(y) dy < \infty,$$

then we have the following equivalent integral inequalities with a homogeneous kernel:

$$\int_0^\infty \int_0^\infty k_\lambda(x,y) f(x) g(y) dx\, dy$$

$$< k_\lambda^{\frac{1}{p}}(\sigma) k_\lambda^{\frac{1}{q}}(\lambda - \mu) \left\{ \int_0^\infty x^{p[1-(\frac{\lambda-\sigma}{p}+\frac{\mu}{q})]-1} f^p(x) dx \right\}^{\frac{1}{p}}$$

$$\times \left\{ \int_0^\infty y^{q[1-(\frac{\sigma}{p}+\frac{\lambda-\mu}{q})]-1} g^q(y) dy \right\}^{\frac{1}{q}}, \tag{2.14}$$

$$\left[\int_0^\infty y^{p(\frac{\sigma}{p}+\frac{\lambda-\mu}{q})-1} \left(\int_0^\infty k_\lambda(x,y) f(x) dx \right)^p dy \right]^{\frac{1}{p}}$$

$$< k_\lambda^{\frac{1}{p}}(\sigma) k_\lambda^{\frac{1}{q}}(\lambda - \mu) \left\{ \int_0^\infty x^{p[1-(\frac{\lambda-\sigma}{p}+\frac{\mu}{q})]-1} f^p(x) dx \right\}^{\frac{1}{p}}. \tag{2.15}$$

Moreover, the constant factor

$$k_\lambda^{\frac{1}{p}}(\sigma)k_\lambda^{\frac{1}{q}}(\lambda - \mu)$$

is the best possible in (2.14) *if and only if the same constant factor in* (2.15) *is the best possible.*

In particular, for $\mu + \sigma = \lambda$, we have the following equivalent inequalities with the best possible constant factor $k_\lambda(\sigma)$:

$$\int_0^\infty \int_0^\infty k_\lambda(x, y)f(x)g(y)dx\, dy$$

$$< k_\lambda(\sigma)\left[\int_0^\infty x^{p(1-\mu)-1}f^p(x)dx\right]^{\frac{1}{p}}$$

$$\times \left[\int_0^\infty y^{q(1-\sigma)-1}g^q(y)dy\right]^{\frac{1}{q}}, \tag{2.16}$$

$$\left[\int_0^\infty y^{p\sigma-1}\left(\int_0^\infty k_\lambda(x, y)f(x)dx\right)^p dy\right]^{\frac{1}{p}}$$

$$< k_\lambda(\sigma)\left[\int_0^\infty x^{p(1-\mu)-1}f^p(x)dx\right]^{\frac{1}{p}}. \tag{2.17}$$

Corollary 2.10. *For $p > 1$, if $k_\lambda(\gamma) \in \mathbf{R}_+$ $(\gamma = \sigma, \lambda - \mu)$, then the following statements,* (I), (II), (III), *and* (IV), *are equivalent:*

(I) both

$$k_\lambda^{\frac{1}{p}}(\sigma)k_\lambda^{\frac{1}{q}}(\lambda - \mu) \quad \text{and} \quad k_\lambda\left(\frac{\sigma}{p} + \frac{\lambda - \mu}{q}\right)$$

are independent of p, q;

(II) $k_\lambda^{\frac{1}{p}}(\sigma)k_\lambda^{\frac{1}{q}}(\lambda - \mu) \leq k_\lambda(\frac{\sigma}{p} + \frac{\lambda-\mu}{q})$;

(III) $\mu + \sigma = \lambda$;

(IV) the constant factor

$$k_\lambda^{\frac{1}{p}}(\sigma)k_\lambda^{\frac{1}{q}}(\lambda - \mu)$$

in (2.14) and (2.15) is the best possible.

Replacing x (resp. y) by x^α (resp. y^β), then replacing $x^{\alpha-1}f(x^\alpha)$ (resp. $y^{\beta-1}g(y^\beta)$) by $f(x)$ (resp. $g(y)$) in Corollary 2.9 and Theorem 2.7, by simplification, we deduce the following.

Corollary 2.11. *For $p > 1$, if $k_\lambda(\gamma) \in \mathbf{R}_+$ $(\gamma = \sigma, \lambda - \mu)$,*

$$0 < \int_0^\infty x^{p[1-\alpha(\frac{\lambda-\sigma}{p}+\frac{\mu}{q})]-1} f^p(x)dx < \infty, \ and$$

$$0 < \int_0^\infty y^{q[1-\beta(\frac{\sigma}{p}+\frac{\lambda-\mu}{q})]-1} g^q(y)dy < \infty,$$

then we have the following equivalent integral inequalities with a homogeneous kernel:

$$\int_0^\infty \int_0^\infty k_\lambda(x^\alpha, y^\beta) f(x) g(y)dx\,dy$$

$$< \frac{1}{\alpha^{1/q}\beta^{1/p}} k_\lambda^{\frac{1}{p}}(\sigma) k_\lambda^{\frac{1}{q}}(\lambda - \mu) \left\{ \int_0^\infty x^{p[1-\alpha(\frac{\lambda-\sigma}{p}+\frac{\mu}{q})]-1} f^p(x)dx \right\}^{\frac{1}{p}}$$

$$\times \left\{ \int_0^\infty y^{q[1-\beta(\frac{\sigma}{p}+\frac{\lambda-\mu}{q})]-1} g^q(y)dy \right\}^{\frac{1}{q}}, \tag{2.18}$$

$$\left[\int_0^\infty y^{p\beta(\frac{\sigma}{p}+\frac{\lambda-\mu}{q})-1} \left(\int_0^\infty k_\lambda(x^\alpha, y^\beta) f(x)dx \right)^p dy \right]^{\frac{1}{p}}$$

$$< \frac{1}{\alpha^{1/q}\beta^{1/p}} k_\lambda^{\frac{1}{p}}(\sigma) k_\lambda^{\frac{1}{q}}(\lambda - \mu)$$

$$\times \left\{ \int_0^\infty x^{p[1-\alpha(\frac{\lambda-\sigma}{p}+\frac{\mu}{q})]-1} f^p(x)dx \right\}^{\frac{1}{p}}. \tag{2.19}$$

Moreover, the constant factor

$$\frac{1}{\alpha^{1/q}\beta^{1/p}} k_\lambda^{\frac{1}{p}}(\sigma) k_\lambda^{\frac{1}{q}}(\lambda - \mu)$$

is the best possible in (2.18) if and only if the same constant factor in (2.19) is the best possible.

In particular, for $\mu + \sigma = \lambda$, we have the following equivalent inequalities with the best possible constant factor

$$\frac{1}{\alpha^{1/q}\beta^{1/p}}k_\lambda(\sigma):$$

$$\int_0^\infty \int_0^\infty k_\lambda(x^\alpha, y^\beta)f(x)g(y)dx\,dy$$

$$< \frac{k_\lambda(\sigma)}{\alpha^{1/q}\beta^{1/p}}\left[\int_0^\infty x^{p(1-\alpha\mu)-1}f^p(x)dx\right]^{\frac{1}{p}}$$

$$\times \left[\int_0^\infty y^{q(1-\beta\sigma)-1}g^q(y)dy\right]^{\frac{1}{q}}, \tag{2.20}$$

$$\left[\int_0^\infty y^{p\beta\sigma-1}\left(\int_0^\infty k_\lambda(x^\alpha, y^\beta)f(x)dx\right)^p dy\right]^{\frac{1}{p}}$$

$$< \frac{k_\lambda(\sigma)}{\alpha^{1/q}\beta^{1/p}}\left\{\int_0^\infty x^{p(1-\alpha\mu)-1}f^p(x)dx\right\}^{\frac{1}{p}}. \tag{2.21}$$

Corollary 2.12. *For $p > 1$,*

$$0 < \int_0^\infty x^{p[1-\alpha(\frac{\sigma}{p}+\frac{\sigma_1}{q})]-1}f^p(x)dx < \infty, \text{ and}$$

$$0 < \int_0^\infty y^{q[1-\beta(\frac{\sigma}{p}+\frac{\sigma_1}{q})]-1}g^q(y)dy < \infty,$$

we then have the following equivalent integral inequalities with a non-homogeneous kernel:

$$\int_0^\infty \int_0^\infty h(x^\alpha y^\beta)f(x)g(y)dx\,dy$$

$$< \frac{1}{\alpha^{1/q}\beta^{1/p}}k^{\frac{1}{p}}(\sigma)k^{\frac{1}{q}}(\sigma_1)\left\{\int_0^\infty x^{p[1-\alpha(\frac{\sigma}{p}+\frac{\sigma_1}{q})]-1}f^p(x)dx\right\}^{\frac{1}{p}}$$

$$\times \left\{\int_0^\infty y^{q[1-\beta(\frac{\sigma}{p}+\frac{\sigma_1}{q})]-1}g^q(y)dy\right\}^{\frac{1}{q}}, \tag{2.22}$$

$$\left[\int_0^\infty y^{p\beta(\frac{\sigma}{p}+\frac{\sigma_1}{q})-1}\left(\int_0^\infty h(x^\alpha y^\beta)f(x)dx\right)^p dy\right]^{\frac{1}{p}}$$

$$< \frac{1}{\alpha^{1/q}\beta^{1/p}}k^{\frac{1}{p}}(\sigma)k^{\frac{1}{q}}(\sigma_1)\left\{\int_0^\infty x^{p[1-\alpha(\frac{\sigma}{p}+\frac{\sigma_1}{q})]-1}f^p(x)dx\right\}^{\frac{1}{p}}.$$

$$(2.23)$$

Moreover, the constant factor

$$\frac{1}{\alpha^{1/q}\beta^{1/p}}k^{\frac{1}{p}}(\sigma)k^{\frac{1}{q}}(\sigma_1)$$

in (2.22) *is the best possible if and only if the same constant factor in* (2.23) *is the best possible.*

In particular, for $\sigma_1 = \sigma$, we have the following equivalent inequalities with the best possible constant factor

$$\frac{1}{\alpha^{1/q}\beta^{1/p}}k(\sigma):$$

$$\int_0^\infty \int_0^\infty h(x^\alpha y^\beta)f(x)g(y)dx\,dy$$

$$< \frac{k(\sigma)}{\alpha^{1/q}\beta^{1/p}}\left[\int_0^\infty x^{p(1-\alpha\sigma)-1}f^p(x)dx\right]^{\frac{1}{p}}$$

$$\times\left[\int_0^\infty y^{q(1-\beta\sigma)-1}g^q(y)dy\right]^{\frac{1}{q}},\qquad(2.24)$$

$$\left[\int_0^\infty y^{p\beta\sigma-1}\left(\int_0^\infty h(x^\alpha y^\beta)f(x)dx\right)^p dy\right]^{\frac{1}{p}}$$

$$< \frac{k(\sigma)}{\alpha^{1/q}\beta^{1/p}}\left[\int_0^\infty x^{p(1-\alpha\sigma)-1}f^p(x)dx\right]^{\frac{1}{p}}.\qquad(2.25)$$

Example 2.13. (i) Setting

$$h(u) = k_\lambda(1,u) = \frac{1}{(1+u)^\lambda}\quad(\lambda > 0),$$

we obtain

$$k(\eta) = k_\lambda(\eta) = \int_0^\infty \frac{u^{\eta-1}du}{(1+u)^\lambda} = B(\eta, \lambda - \eta)\quad(0 < \eta < \lambda).$$

By (2.20) and (2.21), for $p > 1$, $\mu + \sigma = \lambda$, we have the following equivalent extended Hardy–Hilbert integral inequalities with a homogeneous kernel and the best possible constant factor

$$\frac{B(\mu, \sigma)}{\alpha^{1/q}\beta^{1/p}} : \int_0^\infty \int_0^\infty \frac{1}{(x^\alpha + y^\beta)^\lambda} f(x)g(y)dx\,dy$$

$$< \frac{B(\mu, \sigma)}{\alpha^{1/q}\beta^{1/p}} \left[\int_0^\infty x^{p(1-\alpha\mu)-1} f^p(x)dx \right]^{\frac{1}{p}}$$

$$\times \left[\int_0^\infty y^{q(1-\beta\sigma)-1} g^q(y)dy \right]^{\frac{1}{q}}, \tag{2.26}$$

$$\left\{ \int_0^\infty y^{p\beta\sigma-1} \left[\int_0^\infty \frac{1}{(x^\alpha + y^\beta)^\lambda} f(x)dx \right]^p dy \right\}^{\frac{1}{p}}$$

$$< \frac{B(\mu, \sigma)}{\alpha^{1/q}\beta^{1/p}} \left[\int_0^\infty x^{p(1-\alpha\mu)-1} f^p(x)dx \right]^{\frac{1}{p}}. \tag{2.27}$$

In particular, for $\alpha = \beta = 1$, we have

$$\int_0^\infty \int_0^\infty \frac{1}{(x + y)^\lambda} f(x)g(y)dx\,dy$$

$$< B(\mu, \sigma) \left[\int_0^\infty x^{p(1-\mu)-1} f^p(x)dx \right]^{\frac{1}{p}}$$

$$\times \left[\int_0^\infty y^{q(1-\sigma)-1} g^q(y)dy \right]^{\frac{1}{q}}, \tag{2.28}$$

$$\left\{ \int_0^\infty y^{p\sigma-1} \left[\int_0^\infty \frac{1}{(x + y)^\lambda} f(x)dx \right]^p dy \right\}^{\frac{1}{p}}$$

$$< B(\mu, \sigma) \left[\int_0^\infty x^{p(1-\mu)-1} f^p(x)dx \right]^{\frac{1}{p}}. \tag{2.29}$$

By (2.24) and (2.25), for $\sigma_1 = \sigma \in (0, \lambda)$, we have the following equivalent Hardy–Hilbert integral inequalities with a nonhomogeneous kernel and the best possible constant factor

$$\frac{B(\lambda-\sigma,\sigma)}{\alpha^{1/q}\beta^{1/p}} : \int_0^\infty \int_0^\infty \frac{1}{(1+x^\alpha y^\beta)^\lambda} f(x)g(y)dx\,dy$$

$$< \frac{B(\lambda-\sigma,\sigma)}{\alpha^{1/q}\beta^{1/p}} \left[\int_0^\infty x^{p(1-\alpha\sigma)-1} f^p(x)dx\right]^{\frac{1}{p}}$$

$$\times \left[\int_0^\infty y^{q(1-\beta\sigma)-1} g^q(y)dy\right]^{\frac{1}{q}}, \tag{2.30}$$

$$\left\{\int_0^\infty y^{p\beta\sigma-1} \left[\int_0^\infty \frac{1}{(1+x^\alpha y^\beta)^\lambda} f(x)dx\right]^p dy\right\}^{\frac{1}{p}}$$

$$< \frac{B(\lambda-\sigma,\sigma)}{\alpha^{1/q}\beta^{1/p}} \left[\int_0^\infty x^{p(1-\alpha\sigma)-1} f^p(x)dx\right]^{\frac{1}{p}}. \tag{2.31}$$

In particular, for $\alpha = \beta = 1$, we have

$$\int_0^\infty \int_0^\infty \frac{1}{(1+xy)^\lambda} f(x)g(y)dx\,dy$$

$$< B(\lambda-\sigma,\sigma) \left[\int_0^\infty x^{p(1-\sigma)-1} f^p(x)dx\right]^{\frac{1}{p}}$$

$$\times \left[\int_0^\infty y^{q(1-\sigma)-1} g^q(y)dy\right]^{\frac{1}{q}}, \tag{2.32}$$

$$\left\{\int_0^\infty y^{p\sigma-1} \left[\int_0^\infty \frac{1}{(1+xy)^\lambda} f(x)dx\right]^p dy\right\}^{\frac{1}{p}}$$

$$< B(\lambda-\sigma,\sigma) \left[\int_0^\infty x^{p(1-\sigma)-1} f^p(x)dx\right]^{\frac{1}{p}}. \tag{2.33}$$

(ii) Setting

$$h(u) = k_\lambda(1,u) = \frac{\ln u}{u^\lambda - 1} \quad (\lambda > 0),$$

we obtain

$$k(\eta) = k_\lambda(\eta) = \int_0^\infty \frac{u^{\eta-1} \ln u}{u^\lambda - 1} du$$

$$= \left[\frac{\pi}{\lambda \sin(\pi\eta/\lambda)}\right]^2 \quad (0 < \eta < \lambda).$$

By (2.20) and (2.21), for $\mu + \sigma = \lambda$, we have the following equivalent Hilbert-type integral inequalities with a homogeneous kernel and the best possible constant factor

$$\frac{1}{\alpha^{1/q}\beta^{1/p}} \left[\frac{\pi}{\lambda\sin(\pi\sigma/\lambda)}\right]^2 : \int_0^\infty \int_0^\infty \frac{\ln(x^\alpha/y^\beta)}{x^{\alpha\lambda} - y^{\beta\lambda}} f(x)g(y)dx\,dy$$

$$< \frac{1}{\alpha^{1/q}\beta^{1/p}} \left[\frac{\pi}{\lambda\sin(\pi\sigma/\lambda)}\right]^2 \left[\int_0^\infty x^{p(1-\alpha\mu)-1} f^p(x)dx\right]^{\frac{1}{p}}$$

$$\times \left[\int_0^\infty y^{q(1-\beta\sigma)-1} g^q(y)dy\right]^{\frac{1}{q}}, \tag{2.34}$$

$$\left\{\int_0^\infty y^{p\beta\sigma-1} \left[\int_0^\infty \frac{\ln(x^\alpha/y^\beta)}{x^{\alpha\lambda} - y^{\beta\lambda}} f(x)dx\right]^p\right\}^{\frac{1}{p}}$$

$$< \frac{1}{\alpha^{1/q}\beta^{1/p}} \left[\frac{\pi}{\lambda\sin(\pi\sigma/\lambda)}\right]^2 \left[\int_0^\infty x^{p(1-\alpha\mu)-1} f^p(x)dx\right]^{\frac{1}{p}}. \tag{2.35}$$

In particular, for $\alpha = \beta = 1$, it holds that

$$\int_0^\infty \int_0^\infty \frac{\ln(x/y)}{x^\lambda - y^\lambda} f(x)g(y)dx\,dy$$

$$< \left[\frac{\pi}{\lambda\sin(\pi\sigma/\lambda)}\right]^2 \left[\int_0^\infty x^{p(1-\mu)-1} f^p(x)dx\right]^{\frac{1}{p}}$$

$$\times \left[\int_0^\infty y^{q(1-\sigma)-1} g^q(y)dy\right]^{\frac{1}{q}}, \tag{2.36}$$

$$\left\{ \int_0^\infty y^{p\sigma-1} \left[\int_0^\infty \frac{\ln(x/y)}{x^\lambda - y^\lambda} f(x) dx \right]^p dy \right\}^{\frac{1}{p}}$$

$$< \left[\frac{\pi}{\lambda \sin(\pi\sigma/\lambda)} \right]^2 \left[\int_0^\infty x^{p(1-\mu)-1} f^p(x) dx \right]^{\frac{1}{p}}. \quad (2.37)$$

By (2.24) and (2.25), for $\sigma_1 = \sigma \in (0, \lambda)$, we have the following equivalent Hilbert-type integral inequalities with a nonhomogeneous kernel and the best possible constant factor

$$\frac{1}{\alpha^{1/q}\beta^{1/p}} \left[\frac{\pi}{\lambda \sin(\pi\sigma/\lambda)} \right]^2 : \int_0^\infty \int_0^\infty \frac{\ln(x^\alpha y^\beta)}{(x^\alpha y^\beta)^\lambda - 1} f(x)g(y) dx \, dy$$

$$< \frac{1}{\alpha^{1/q}\beta^{1/p}} \left[\frac{\pi}{\lambda \sin(\pi\sigma/\lambda)} \right]^2 \left[\int_0^\infty x^{p(1-\alpha\sigma)-1} f^p(x) dx \right]^{\frac{1}{p}}$$

$$\times \left[\int_0^\infty y^{q(1-\beta\sigma)-1} g^q(y) dy \right]^{\frac{1}{q}}, \quad (2.38)$$

$$\left\{ \int_0^\infty y^{p\beta\sigma-1} \left[\int_0^\infty \frac{\ln(x^\alpha y^\beta)}{(x^\alpha y^\beta)^\lambda - 1} f(x) dx \right]^p dy \right\}^{\frac{1}{p}}$$

$$< \frac{1}{\alpha^{1/q}\beta^{1/p}} \left[\frac{\pi}{\lambda \sin(\pi\sigma/\lambda)} \right]^2 \left[\int_0^\infty x^{p(1-\alpha\sigma)-1} f^p(x) dx \right]^{\frac{1}{p}}.$$

$$(2.39)$$

In particular, for $\alpha = \beta = 1$, we have

$$\int_0^\infty \int_0^\infty \frac{\ln(xy)}{(xy)^\lambda - 1} f(x)g(y) dx \, dy$$

$$< \left[\frac{\pi}{\lambda \sin(\pi\sigma/\lambda)} \right]^2 \left[\int_0^\infty x^{p(1-\sigma)-1} f^p(x) dx \right]^{\frac{1}{p}}$$

$$\times \left[\int_0^\infty y^{q(1-\sigma)-1} g^q(y) dy \right]^{\frac{1}{q}}, \quad (2.40)$$

$$\left\{ \int_0^\infty y^{p\sigma-1} \left[\int_0^\infty \frac{\ln(xy)}{(xy)^\lambda - 1} f(x) dx \right]^p dy \right\}^{\frac{1}{p}}$$

$$< \left[\frac{\pi}{\lambda \sin(\pi\sigma/\lambda)} \right]^2 \left[\int_0^\infty x^{p(1-\sigma)-1} f^p(x) dx \right]^{\frac{1}{p}}. \quad (2.41)$$

(iii) Setting

$$h(u) = k_\lambda(1, u) = \frac{1}{(\max\{1, u\})^\lambda} \quad (\lambda > 0),$$

we obtain

$$k(\eta) = k_\lambda(\eta) = \int_0^\infty \frac{u^{\eta-1} du}{(\max\{1, u\})^\lambda} = \frac{\lambda}{\eta(\lambda - \eta)} \quad (0 < \eta < \lambda).$$

By (2.20) and (2.21), for $p > 1$, and $\mu + \sigma = \lambda$, we have the following equivalent extended Hardy–Littlewood–Polya integral inequalities with a homogeneous kernel and the best possible constant factor

$$\frac{\lambda}{\alpha^{1/q}\beta^{1/p}\mu\sigma} : \int_0^\infty \int_0^\infty \frac{1}{(\max\{x^\alpha, y^\beta\})^\lambda} f(x)g(y)dx\,dy$$

$$< \frac{\lambda}{\alpha^{1/q}\beta^{1/p}\mu\sigma} \left[\int_0^\infty x^{p(1-\alpha\mu)-1} f^p(x)dx\right]^{\frac{1}{p}}$$

$$\times \left[\int_0^\infty y^{q(1-\beta\sigma)-1} g^q(y)dy\right]^{\frac{1}{q}}, \tag{2.42}$$

$$\left\{\int_0^\infty y^{p\beta\sigma-1}\left[\int_0^\infty \frac{1}{(\max\{x^\alpha, y^\beta\})^\lambda} f(x)dx\right]^p dy\right\}^{\frac{1}{p}}$$

$$< \frac{\lambda}{\alpha^{1/q}\beta^{1/p}\mu\sigma} \left[\int_0^\infty x^{p(1-\alpha\mu)-1} f^p(x)dx\right]^{\frac{1}{p}}. \tag{2.43}$$

In particular, for $\alpha = \beta = 1$, we obtain

$$\int_0^\infty \int_0^\infty \frac{1}{(\max\{x, y\})^\lambda} f(x)g(y)dx\,dy$$

$$< \frac{\lambda}{\mu\sigma} \left[\int_0^\infty x^{p(1-\mu)-1} f^p(x)dx\right]^{\frac{1}{p}}$$

$$\times \left[\int_0^\infty y^{q(1-\sigma)-1} g^q(y)dy\right]^{\frac{1}{q}}, \tag{2.44}$$

$$\left\{ \int_0^\infty y^{p\sigma-1} \left[\int_0^\infty \frac{1}{(\max\{x,y\})^\lambda} f(x)dx \right]^p dy \right\}^{\frac{1}{p}}$$

$$< \frac{\lambda}{\mu\sigma} \left[\int_0^\infty x^{p(1-\mu)-1} f^p(x)dx \right]^{\frac{1}{p}}. \tag{2.45}$$

By (2.24) and (2.25), for $\sigma_1 = \sigma \in (0,\lambda)$, we have the following equivalent extended Hardy–Littlewood–Polya integral inequalities with a nonhomogeneous kernel and the best possible constant factor

$$\frac{\lambda}{\alpha^{1/q}\beta^{1/p}(\lambda-\sigma)\sigma} : \int_0^\infty \int_0^\infty \frac{1}{(\max\{1,x^\alpha y^\beta\})^\lambda} f(x)g(y)dx\,dy$$

$$< \frac{\lambda}{\alpha^{1/q}\beta^{1/p}(\lambda-\sigma)\sigma} \left[\int_0^\infty x^{p(1-\alpha\sigma)-1} f^p(x)dx \right]^{\frac{1}{p}}$$

$$\times \left[\int_0^\infty y^{q(1-\beta\sigma)-1} g^q(y)dy \right]^{\frac{1}{q}}, \tag{2.46}$$

$$\left\{ \int_0^\infty y^{p\beta\sigma-1} \left[\int_0^\infty \frac{1}{(\max\{1,x^\alpha y^\beta\})^\lambda} f(x)dx \right]^p dy \right\}^{\frac{1}{p}}$$

$$< \frac{\lambda}{\alpha^{1/q}\beta^{1/p}(\lambda-\sigma)\sigma} \left[\int_0^\infty x^{p(1-\alpha\sigma)-1} f^p(x)dx \right]^{\frac{1}{p}}. \tag{2.47}$$

In particular, for $\alpha = \beta = 1$, we have

$$\int_0^\infty \int_0^\infty \frac{1}{(\max\{1,xy\})^\lambda} f(x)g(y)dx\,dy$$

$$< \frac{\lambda}{(\lambda-\sigma)\sigma} \left[\int_0^\infty x^{p(1-\sigma)-1} f^p(x)dx \right]^{\frac{1}{p}}$$

$$\times \left[\int_0^\infty y^{q(1-\sigma)-1} g^q(y)dy \right]^{\frac{1}{q}}, \tag{2.48}$$

$$\left\{ \int_0^\infty y^{p\sigma-1} \left[\int_0^\infty \frac{1}{(\max\{1, xy\})^\lambda} f(x)dx \right]^p dy \right\}^{\frac{1}{p}}$$

$$< \frac{\lambda}{(\lambda - \sigma)\sigma} \left[\int_0^\infty x^{p(1-\sigma)-1} f^p(x)dx \right]^{\frac{1}{p}}. \qquad (2.49)$$

(iii) For $b \in (0, 1)$, we have

$$
\begin{aligned}
A : &= \int_0^\infty \frac{u^{b-1}}{1-u} du = \int_0^1 \frac{u^{b-1}}{1-u} du + \int_1^\infty \frac{u^{b-1}}{1-u} du \\
&= \int_0^1 \frac{u^{b-1}}{1-u} du - \int_0^1 \frac{v^{-b}}{1-v} dv = \int_0^1 \frac{u^{b-1} - u^{-b}}{1-u} du \\
&= \int_0^1 \sum_{k=0}^\infty (u^{k+b-1} - u^{k-b}) du = \sum_{k=0}^\infty \int_0^1 (u^{k+b-1} - u^{k-b}) du \\
&= \sum_{k=0}^\infty \left(\frac{1}{k+b} - \frac{1}{k+1-b} \right) \\
&= \pi \left[\frac{1}{\pi b} + \sum_{k=1}^\infty \left(\frac{1}{\pi b - \pi k} + \frac{1}{\pi b + \pi k} \right) \right].
\end{aligned}
$$

Since (cf. Ref. [121])

$$\cot x = \frac{1}{x} + \sum_{k=1}^\infty \left(\frac{1}{x - \pi k} + \frac{1}{x + \pi k} \right) \ (x \in (0, \pi)),$$

we obtain that

$$A = \pi \cot(\pi b).$$

Setting

$$h(u) = k_\lambda(1, u) = \frac{1 - u^\eta}{1 - u^{\lambda+\eta}} \quad (\lambda, \eta > 0),$$

we obtain for $\sigma, \mu \in (0, \lambda)$, $\mu + \sigma = \lambda$,

$$
\begin{aligned}
k(\sigma) \quad = \quad & k_\lambda(\sigma) = \int_0^\infty \frac{1 - u^\eta}{1 - u^{\lambda+\eta}} u^{\sigma-1} du \\
\overset{v=u^{\lambda+\eta}}{=} \quad & \frac{1}{\lambda+\eta} \int_0^\infty \frac{1 - v^{\frac{\eta}{\lambda+\eta}}}{1 - v} v^{\frac{\sigma}{\lambda+\eta}-1} dv \\
= \quad & \frac{1}{\lambda+\eta} \left(\int_0^\infty \frac{v^{\frac{\sigma}{\lambda+\eta}-1}}{1 - v} dv - \int_0^\infty \frac{v^{\frac{\eta+\sigma}{\lambda+\eta}-1}}{1 - v} dv \right) \\
= \quad & \frac{\pi}{\lambda+\eta} \left[\cot(\frac{\pi\sigma}{\lambda+\eta}) - \cot(\frac{\pi(\eta+\sigma)}{\lambda+\eta}) \right] \\
= \quad & \frac{\pi}{\lambda+\eta} \left[\cot(\frac{\pi\sigma}{\lambda+\eta}) + \cot(\frac{\pi\mu}{\lambda+\eta}) \right] \in \mathbf{R}_+ .
\end{aligned}
$$

By (2.20) and (2.21), for $p > 1$, $\mu + \sigma = \lambda$, we have the following equivalent Hilbert-type integral inequalities with a homogeneous kernel and the best possible constant factor

$$
\frac{k_\lambda(\sigma)}{\alpha^{1/q}\beta^{1/p}} : \int_0^\infty \int_0^\infty \frac{x^{\alpha\eta} - y^{\beta\eta}}{x^{\alpha(\lambda+\eta)} - y^{\beta(\lambda+\eta)}} f(x)g(y) dx\, dy
$$

$$
< \frac{k_\lambda(\sigma)}{\alpha^{1/q}\beta^{1/p}} \left[\int_0^\infty x^{p(1-\alpha\mu)-1} f^p(x) dx \right]^{\frac{1}{p}}
$$

$$
\times \left[\int_0^\infty y^{q(1-\beta\sigma)-1} g^q(y) dy \right]^{\frac{1}{q}}, \tag{2.50}
$$

$$
\left\{ \int_0^\infty y^{p\beta\sigma-1} \left[\int_0^\infty \frac{x^{\alpha\eta} - y^{\beta\eta}}{x^{\alpha(\lambda+\eta)} - y^{\beta(\lambda+\eta)}} f(x) dx \right]^p dy \right\}^{\frac{1}{p}}
$$

$$
< \frac{k_\lambda(\sigma)}{\alpha^{1/q}\beta^{1/p}} \left[\int_0^\infty x^{p(1-\alpha\mu)-1} f^p(x) dx \right]^{\frac{1}{p}} . \tag{2.51}
$$

By (2.24) and (2.25), for $\sigma_1 = \sigma$, we have the following equivalent Hilbert-type integral inequalities with a nonhomogeneous kernel and the best possible constant factor

$$\frac{k(\sigma)}{\alpha^{1/q}\beta^{1/p}} : \int_0^\infty \int_0^\infty \frac{1-(x^\alpha y^\beta)^\eta}{1-(x^\alpha y^\beta)^{\lambda+\eta}} f(x)g(y)dx\,dy$$

$$< \frac{k(\sigma)}{\alpha^{1/q}\beta^{1/p}} \left[\int_0^\infty x^{p(1-\alpha\sigma)-1} f^p(x)dx \right]^{\frac{1}{p}}$$

$$\times \left[\int_0^\infty y^{q(1-\beta\sigma)-1} g^q(y)dy \right]^{\frac{1}{q}}, \tag{2.52}$$

$$\left\{ \int_0^\infty y^{p\beta\sigma-1} \left[\int_0^\infty \frac{1-(x^\alpha y^\beta)^\eta}{1-(x^\alpha y^\beta)^{\lambda+\eta}} f(x)dx \right]^p dy \right\}^{\frac{1}{p}}$$

$$< \frac{k(\sigma)}{\alpha^{1/q}\beta^{1/p}} \left[\int_0^\infty x^{p(1-\alpha\sigma)-1} f^p(x)dx \right]^{\frac{1}{p}}. \tag{2.53}$$

2.3   Operator Expressions and Applications

(a) For $p > 1$, we set

$$\varphi(x) := x^{p[1-(\frac{\sigma}{p}+\frac{\sigma_1}{q})]-1} \quad \text{and} \quad \psi(y) := y^{q[1-(\frac{\sigma}{p}+\frac{\sigma_1}{q})]-1},$$

wherefrom we derive that

$$\psi^{1-p}(y) = y^{p(\frac{\sigma}{p}+\frac{\sigma_1}{q})-1} \quad (x, y \in \mathbf{R}_+).$$

Define some real normed linear spaces as follows:

$$L_{p,\varphi}(\mathbf{R}_+) := \left\{ f = f(x) : \|f\|_{p,\varphi} \right.$$

$$:= \left(\int_0^\infty \varphi(x)|f(x)|^p dx \right)^{\frac{1}{p}} < \infty \right\},$$

$$L_{q,\psi}(\mathbf{R}_+) = \left\{ g = g(y) : \|g\|_{q,\psi} \right.$$

$$:= \left(\int_0^\infty \psi(y)|g(y)|^q dy \right)^{\frac{1}{q}} < \infty \right\},$$

$$L_{p,\psi^{1-p}}(\mathbf{R}_+) = \left\{ h = h(y) : ||h||_{p,\psi^{1-p}} \right.$$

$$= \left. \left(\int_0^\infty \psi^{1-p}(y)|h(y)|^p dy \right)^{\frac{1}{p}} < \infty \right\}.$$

For $f \in L_{p,\varphi}(\mathbf{R}_+)$, setting

$$H(y) := \int_0^\infty h(xy)f(x)dx \ (y \in \mathbf{R}_+),$$

we may rewrite (2.10) in the following manner:

$$||H||_{p,\psi^{1-p}} = \left[\int_0^\infty \psi^{1-p}(y)H^p(y)dy \right]^{\frac{1}{p}} < k^{\frac{1}{p}}(\sigma)k^{\frac{1}{q}}(\sigma_1)||f||_{p,\varphi} < \infty,$$

$$(2.54)$$

namely,

$$H \in L_{p,\psi^{1-p}}(\mathbf{R}_+).$$

Definition 2.14. Define a Hilbert-type integral operator with the nonhomogeneous kernel

$$T : L_{p,\varphi}(\mathbf{R}_+) \to L_{p,\psi^{1-p}}(\mathbf{R}_+)$$

as follows.

For any $f \in L_{p,\varphi}(\mathbf{R}_+)$, there exists a unique representation

$$H = Tf \in L_{p,\psi^{1-p}}(\mathbf{R}_+)$$

satisfying

$$Tf(y) = H(y)$$

for any $y \in \mathbf{R}_+$.

Define the formal inner product of Tf and $g \in L_{q,\psi}(\mathbf{R}_+)$ and the norm of T as follows:

$$(Tf, g) := \int_0^\infty \left(\int_0^\infty h(xy)f(x)dx \right) g(y)dy = I,$$

$$\|T\| = \sup_{f(\neq \theta) \in L_{p,\varphi}(\mathbf{R}_+)} \frac{\|Tf\|_{p,\psi^{1-p}}}{\|f\|_{p,\varphi}}.$$

In view of (2.54), it follows that

$$\|Tf\|_{p,\psi^{1-p}} = \|H\|_{p,\psi^{1-p}} \le k^{\frac{1}{p}}(\sigma)k^{\frac{1}{q}}(\sigma_1)\|f\|_{p,\varphi},$$

and then, the operator T is bounded, satisfying

$$\|T\| \le k^{\frac{1}{p}}(\sigma)k^{\frac{1}{q}}(\sigma_1).$$

By Theorems 2.7 and 2.8, we deduce the following theorem.

Theorem 2.15. *If $f(> 0) \in L_{p,\varphi}(\mathbf{R}_+), g(> 0) \in L_{q,\psi}(\mathbf{R}_+)$, then we have the following equivalent inequalities:*

$$(Tf, g) < k^{\frac{1}{p}}(\sigma)k^{\frac{1}{q}}(\sigma_1)\|f\|_{p,\varphi}\|g\|_{q,\psi}, \tag{2.55}$$

$$\|Tf\|_{p,\psi^{1-p}} < k^{\frac{1}{p}}(\sigma)k^{\frac{1}{q}}(\sigma_1)\|f\|_{p,\varphi}. \tag{2.56}$$

Moreover, $\sigma_1 = \sigma$ if and only if the constant factor $k^{\frac{1}{p}}(\sigma)k^{\frac{1}{q}}(\sigma_1)$ in (2.55) and (2.56) is the best possible, namely,

$$\|T\| = k(\sigma).$$

(b) For $p > 1$, we set

$$\Phi(x) := x^{p[1-(\frac{\lambda-\sigma}{p}+\frac{\mu}{q})]-1} \quad \text{and} \quad \Psi(y) := y^{q[1-(\frac{\sigma}{p}+\frac{\lambda-\mu}{q})]-1},$$

wherefrom

$$\Psi^{1-p}(y) = y^{p(\frac{\sigma}{p}+\frac{\lambda-\mu}{q})-1}(x, y \in \mathbf{R}_+).$$

Define some real normed linear spaces as follows:

$$L_{p,\Phi}(\mathbf{R}_+) := \left\{ f = f(x) : ||f||_{p,\Phi} \right.$$

$$:= \left(\int_0^\infty \Phi(x)|f(x)|^p dx \right)^{\frac{1}{p}} < \infty \left. \right\},$$

$$L_{q,\Psi}(\mathbf{R}_+) = \left\{ g = g(y) : ||g||_{q,\Psi} \right.$$

$$:= \left(\int_0^\infty \Psi(y)|g(y)|^q dy \right)^{\frac{1}{q}} < \infty \left. \right\},$$

$$L_{p,\Psi^{1-p}}(\mathbf{R}_+) = \left\{ h = h(y) : ||h||_{p,\Psi^{1-p}} \right.$$

$$= \left(\int_0^\infty \Psi^{1-p}(y)|h(y)|^p dy \right)^{\frac{1}{p}} < \infty \left. \right\}.$$

For $f \in L_{p,\Phi}(\mathbf{R}_+)$, setting

$$H_1(y) := \int_0^\infty k_\lambda(x,y) f(x) dx \ (y \in \mathbf{R}_+),$$

we may rewrite (2.15) as follows:

$$||H_1||_{p,\Psi^{1-p}} = \left[\int_0^\infty \Psi^{1-p}(y) H_1^p(y) dy \right]^{\frac{1}{p}}$$

$$< k_\lambda^{\frac{1}{p}}(\sigma) k_\lambda^{\frac{1}{q}}(\lambda - \mu) ||f||_{p,\Phi} < \infty, \qquad (2.57)$$

namely, $H_1 \in L_{p,\Psi^{1-p}}(\mathbf{R}_+)$.

Definition 2.16. Define a Hilbert-type integral operator with the homogeneous kernel

$$T_1 : L_{p,\Phi}(\mathbf{R}_+) \to L_{p,\Psi^{1-p}}(\mathbf{R}_+)$$

as follows.

For any $f \in L_{p,\Phi}(\mathbf{R}_+)$, there exists a unique representation

$$H_1 = T_1 f \in L_{p,\Psi^{1-p}}(\mathbf{R}_+),$$

satisfying

$$T_1 f(y) = H_1(y)$$

for any $y \in \mathbf{R}_+$.

Define the formal inner product of $T_1 f$ and $g \in L_{q,\Psi}(\mathbf{R}_+)$ and the norm of T_1 as follows:

$$(T_1 f, g) \ : \ = \int_0^\infty \left(\int_0^\infty k_\lambda(x, y) f(x) dx \right) g(y) dy,$$

$$||T_1|| = \sup_{f(\neq\theta)\in L_{p,\Phi}(\mathbf{R}_+)} \frac{||T_1 f||_{p,\Psi^{1-p}}}{||f||_{p,\Phi}}.$$

In view of (2.57), it follows that

$$||T_1 f||_{p,\Psi^{1-p}} = ||H_1||_{p,\Psi^{1-p}} \leq k_\lambda^{\frac{1}{p}}(\sigma) k_\lambda^{\frac{1}{q}}(\lambda - \mu) ||f||_{p,\Phi},$$

and then, the operator T_1 is bounded, satisfying

$$||T_1|| \leq k_\lambda^{\frac{1}{p}}(\sigma) k_\lambda^{\frac{1}{q}}(\lambda - \mu).$$

By Corollaries 2.9 and 2.10, we deduce the following corollary.

Corollary 2.17. *For $p > 1$, if $f(>0) \in L_{p,\Phi}(\mathbf{R}_+), g(>0) \in L_{q,\Psi}(\mathbf{R}_+)$, then we have the following equivalent inequalities:*

$$(T_1 f, g) < k_\lambda^{\frac{1}{p}}(\sigma) k_\lambda^{\frac{1}{q}}(\lambda - \mu) ||f||_{p,\Phi} ||g||_{q,\Psi}, \qquad (2.58)$$

$$||T_1 f||_{p,\Psi^{1-p}} < k_\lambda^{\frac{1}{p}}(\sigma) k_\lambda^{\frac{1}{q}}(\lambda - \mu) ||f||_{p,\Phi}. \qquad (2.59)$$

Moreover, $\mu + \sigma = \lambda$ if and only if the constant factor

$$k_\lambda^{\frac{1}{p}}(\sigma) k_\lambda^{\frac{1}{q}}(\lambda - \mu)$$

in (2.58) and (2.59) is the best possible, namely,

$$||T_1|| = k_\lambda(\sigma).$$

Example 2.18. (i) Setting

$$h(u) = k_\lambda(1, u) = \frac{(\min\{1, u\})^\gamma |\ln u|^\beta}{|u^{\lambda+\gamma} - 1|} \quad (\beta > 0, \lambda > -\gamma),$$

we derive that

$$h(xy) = \frac{(\min\{1, xy\})^\gamma |\ln(xy)|^\beta}{|(xy)^{\lambda+\gamma} - 1|} \quad \text{and}$$

$$k_\lambda(x, y) = \frac{(\min\{x, y\})^\gamma |\ln(x/y)|^\beta}{|x^{\lambda+\gamma} - y^{\lambda+\gamma}|}.$$

For $\sigma, \mu = \lambda - \sigma \in (-\gamma, \lambda)$, we obtain

$$k(\sigma) = k_\lambda(\sigma) = \int_0^\infty \frac{(\min\{1, u\})^\gamma |\ln u|^\beta}{|u^{\lambda+\gamma} - 1|} u^{\sigma-1} du$$

$$= \int_0^1 \frac{u^\gamma (-\ln u)^\beta}{1 - u^{\lambda+\gamma}} u^{\sigma-1} du + \int_1^\infty \frac{(\ln u)^\beta}{u^{\lambda+\gamma} - 1} u^{\sigma-1} du$$

$$= \int_0^1 \frac{(-\ln u)^\beta}{1 - u^{\lambda+\gamma}} (u^{\sigma+\gamma-1} + u^{\mu+\gamma-1}) du$$

$$= \int_0^1 (-\ln u)^\beta \sum_{k=0}^\infty u^{k(\lambda+\gamma)} (u^{\sigma+\gamma-1} + u^{\mu+\gamma-1}) du.$$

By Lebesgue's term-by-term integration theorem (cf. Ref. [119]), we have

$$k(\sigma) = k_\lambda(\sigma) = \sum_{k=0}^\infty \int_0^1 (-\ln u)^\beta [u^{k(\lambda+\gamma)+\sigma+\gamma-1} + u^{k(\lambda+\gamma)+\mu+\gamma-1}] du$$

$$= \int_0^\infty v^\beta e^{-v} dv \sum_{k=0}^\infty \left\{ \frac{1}{[k(\lambda+\gamma) + \sigma + \gamma]^{\beta+1}} \right.$$

$$\left. + \frac{1}{[k(\lambda+\gamma) + \mu + \gamma]^{\beta+1}} \right\}$$

$$= \frac{\Gamma(\beta+1)}{(\lambda+\gamma)^{\beta+1}} \left(\zeta\left(\beta+1, \frac{\sigma+\gamma}{\lambda+\gamma}\right) + \zeta\left(\beta+1, \frac{\mu+\gamma}{\lambda+\gamma}\right) \right),$$

where

$$\zeta(s,a) := \sum_{k=0}^{\infty} \frac{1}{(k+a)^s} \quad (\text{Re}\, s > 1, 0 < a \le 1)$$

is the Hurwitz zeta function (with $\zeta(s,1) = \zeta(s)$ being the Riemann zeta function; cf. Refs. [123–125]).

Then, by Theorem 2.15 and Corollary 2.17, we have

$$\|T\| = \|T_1\| = \frac{\Gamma(\beta+1)}{(\lambda+\gamma)^{\beta+1}} \left(\zeta\left(\beta+1, \frac{\sigma+\gamma}{\lambda+\gamma}\right) \right.$$
$$\left. + \zeta\left(\beta+1, \frac{\mu+\gamma}{\lambda+\gamma}\right) \right).$$

(ii) Setting

$$h(u) = k_\lambda(1,u) = \frac{(\max\{1,u\})^\gamma |\ln u|^\beta}{|u^{\lambda+\gamma} - 1|} \quad (\beta > 0, \lambda > -\gamma),$$

we obtain that

$$h(xy) = \frac{(\max\{1,xy\})^\gamma |\ln(xy)|^\beta}{|(xy)^{\lambda+\gamma} - 1|} \quad \text{and}$$

$$k_\lambda(x,y) = \frac{(\max\{x,y\})^\gamma |\ln(x/y)|^\beta}{|x^{\lambda+\gamma} - y^{\lambda+\gamma}|}.$$

For $\sigma, \mu = \lambda - \sigma \in (0, \lambda + \gamma)$, it follows that

$$k(\sigma) = k_\lambda(\sigma) = \int_0^\infty \frac{(\max\{1,u\})^\gamma |\ln u|^\beta}{|u^{\lambda+\gamma} - 1|} u^{\sigma-1} du$$

$$= \int_0^1 \frac{(-\ln u)^\beta}{1 - u^{\lambda+\gamma}} u^{\sigma-1} du + \int_1^\infty \frac{u^\gamma (\ln u)^\beta}{u^{\lambda+\gamma} - 1} u^{\sigma-1} du$$

$$= \int_0^1 \frac{(-\ln u)^\beta}{1 - u^{\lambda+\gamma}} (u^{\sigma-1} + u^{\mu-1}) du$$

$$= \int_0^1 (-\ln u)^\beta \sum_{k=0}^{\infty} u^{k(\lambda+\gamma)} (u^{\sigma-1} + u^{\mu-1}) du.$$

By Lebesgue's term-by-term integration theorem (cf. Ref. [119]), we have

$$k(\sigma) = k_\lambda(\sigma) = \sum_{k=0}^{\infty} \int_0^1 (-\ln u)^\beta [u^{k(\lambda+\gamma)+\sigma-1} + u^{k(\lambda+\gamma)+\mu-1}] du$$

$$= \int_0^\infty v^\beta e^{-v} dv \sum_{k=0}^{\infty} \left\{ \frac{1}{[k(\lambda+\gamma)+\sigma]^{\beta+1}} + \frac{1}{[k(\lambda+\gamma)+\mu]^{\beta+1}} \right\}$$

$$= \frac{\Gamma(\beta+1)}{(\lambda+\gamma)^{\beta+1}} \left(\zeta\left(\beta+1, \frac{\sigma}{\lambda+\gamma}\right) + \zeta\left(\beta+1, \frac{\mu}{\lambda+\gamma}\right) \right).$$

Then, by Theorem 2.15 and Corollary 2.17, we have

$$\|T\| = \|T_1\| = \frac{\Gamma(\beta+1)}{(\lambda+\gamma)^{\beta+1}} \left(\zeta\left(\beta+1, \frac{\sigma}{\lambda+\gamma}\right) + \zeta\left(\beta+1, \frac{\mu}{\lambda+\gamma}\right) \right).$$

Example 2.19. (i) Setting

$$h(u) = k_0(1, u) = \cot h \, u - 1 = \frac{2e^{-u}}{e^u - e^{-u}},$$

where

$$\cot h \, u := \frac{e^u + e^{-u}}{e^u - e^{-u}} (u \in \mathbf{R}_+)$$

is a hyperbolic cotangent function (cf. Ref. [122]), we derive that

$$h(xy) = \cot h \, (xy) - 1 = \frac{2e^{-xy}}{e^{xy} - e^{-xy}} \quad \text{and}$$

$$k_0(x, y) = \cot h \left(\frac{y}{x}\right) - 1 = \frac{2e^{-y/x}}{e^{y/x} - e^{-y/x}}.$$

For $\sigma \in (1, \infty)$, we obtain

$$k(\sigma) = k_0(\sigma) = \int_0^\infty \frac{2e^{-u} u^{\sigma-1}}{e^u - e^{-u}} du = \int_0^\infty \frac{2e^{-2u} u^{\sigma-1}}{1 - e^{-2u}} du$$

$$= 2 \int_0^\infty u^{\sigma-1} \sum_{k=0}^{\infty} e^{-2(k+1)u} du.$$

By Lebesgue's term-by-term integration theorem (cf. Ref. [119]), we have

$$k(\sigma) = k_0(\sigma) = 2 \sum_{k=0}^{\infty} \int_0^{\infty} u^{\sigma-1} e^{-2(k+1)u} du$$

$$= 2 \sum_{k=0}^{\infty} \frac{1}{[2(k+1)]^{\sigma}} \int_0^{\infty} v^{\sigma-1} e^{-v} dv = 2^{1-\sigma} \Gamma(\sigma) \zeta(\sigma).$$

Then, by Theorem 2.15 and Corollary 2.12, we have

$$\|T\| = \|T_1\| = 2^{1-\sigma} \Gamma(\sigma) \zeta(\sigma).$$

(ii) Setting

$$h(u) = k_0(1, u) = 1 - \tan h\, u = \frac{2e^{-u}}{e^u + e^{-u}},$$

where

$$\tan h\, u := \frac{e^u - e^{-u}}{e^u + e^{-u}} (u \in \mathbf{R}_+)$$

is a hyperbolic tangent function (cf. Ref. [122]), we obtain that

$$h(xy) = 1 - \tan h\,(xy) = \frac{2e^{-xy}}{e^{xy} + e^{-xy}} \quad \text{and}$$

$$k_0(x, y) = 1 - \tan h\left(\frac{y}{x}\right) = \frac{2e^{-y/x}}{e^{y/x} + e^{-y/x}}.$$

For $\sigma \in (1, \infty)$, we deduce that

$$k(\sigma) = k_0(\sigma) = \int_0^{\infty} \frac{2e^{-u} u^{\sigma-1}}{e^u + e^{-u}} du = \int_0^{\infty} \frac{2e^{-2u} u^{\sigma-1}}{1 + e^{-2u}} du$$

$$= 2 \int_0^{\infty} u^{\sigma-1} \sum_{k=0}^{\infty} (-1)^k e^{-2(k+1)u} du$$

$$= 2 \int_0^{\infty} u^{\sigma-1} \sum_{k=0}^{\infty} \left[e^{-2(2k+1)u} - e^{-4(k+1)u} \right] du.$$

By Lebesgue's term-by-term integration theorem (cf. Ref. [119]), we have

$$k(\sigma) = k_0(\sigma) = 2 \sum_{k=0}^{\infty} \int_0^{\infty} u^{\sigma-1} \left[e^{-2(2k+1)u} - e^{-4(k+1)u} \right] du$$

$$= 2^{1-\sigma} \Gamma(\sigma) \sum_{k=0}^{\infty} \left[\frac{1}{(2k+1)^{\sigma}} - \frac{1}{[2(k+1)]^{\sigma}} \right]$$

$$= 2^{1-\sigma}(1 - 2^{1-\sigma})\Gamma(\sigma) \sum_{k=0}^{\infty} \frac{1}{(k+1)^{\sigma}}$$

$$= 2^{1-\sigma}(1 - 2^{1-\sigma})\Gamma(\sigma)\zeta(\sigma).$$

Then, by Theorem 2.15 and Corollary 2.17, we obtain that

$$\|T\| = \|T_1\| = 2^{1-\sigma}(1 - 2^{1-\sigma})\Gamma(\sigma)\zeta(\sigma).$$

2.4   The Case of Reverses

Lemma 2.20. *For $0 < p < 1$ $(q < 0)$, if there exists a constant $\delta_0 > 0$ such that $k(\sigma \pm \delta_0) < \infty$, then the constant factor in the reverse of (2.7) is the best possible.*

Proof.   For any $0 < \varepsilon < p\delta_0$, we set

$$\widetilde{f}(x) := \begin{cases} 0, & 0 < x < 1, \\ x^{\sigma - \frac{\varepsilon}{p} - 1}, & x \geq 1; \end{cases} \qquad \widetilde{g}(y) := \begin{cases} y^{\sigma + \frac{\varepsilon}{q} - 1}, & 0 < y \leq 1, \\ 0, & y > 1. \end{cases}$$

If there exists a positive constant M $(\geq k(\sigma))$ such that the reverse of (2.7) is valid when replacing $k(\sigma)$ by M, then in particular, we have

$$\widetilde{I} := \int_0^{\infty} \int_0^{\infty} h(xy)\widetilde{f}(x)\widetilde{g}(y)dx\,dy$$

$$> M \left[\int_0^{\infty} x^{p(1-\sigma)-1} \widetilde{f}^p(x)dx \right]^{\frac{1}{p}} \left[\int_0^{\infty} y^{q(1-\sigma)-1} \widetilde{g}^q(y)dy \right]^{\frac{1}{q}}$$

$$= M \left(\int_1^{\infty} x^{-\varepsilon-1}dx \right)^{\frac{1}{p}} \left(\int_0^1 y^{\varepsilon-1}dy \right)^{\frac{1}{q}} = \frac{M}{\varepsilon}.$$

By (2.5), in view of $\sigma - \frac{\varepsilon}{p} \in (\sigma - \delta_0, \sigma + \delta_0)$, we obtain that

$$
\begin{aligned}
\tilde{I} &= \int_0^1 \left(\int_1^\infty h(xy) x^{\sigma - \frac{\varepsilon}{p} - 1} dx \right) y^{\sigma + \frac{\varepsilon}{q} - 1} dy \\
&\leq \int_0^1 \left(\int_0^\infty h(xy) x^{\sigma - \frac{\varepsilon}{p} - 1} dx \right) y^{\sigma + \frac{\varepsilon}{q} - 1} dy \\
&= \int_0^1 \omega \left(\sigma - \frac{\varepsilon}{p}, y \right) y^{\varepsilon - 1} dy = k \left(\sigma - \frac{\varepsilon}{p} \right) \int_0^1 y^{\varepsilon - 1} dy \\
&= \frac{1}{\varepsilon} k \left(\sigma - \frac{\varepsilon}{p} \right) < \infty.
\end{aligned}
$$

In view of the above results, it follows that

$$
\infty > k \left(\sigma - \frac{\varepsilon}{p} \right) \geq \varepsilon \tilde{I} > M.
$$

For $\varepsilon \to 0^+$, by Lemma 2.1, we derive that $k(\sigma) \geq M$. Hence, $M = k(\sigma)$ is the best possible constant factor in the reverse of (2.7). This completes the proof of the lemma. $\qquad\square$

Remark 2.21. For $0 < p < 1$, we set

$$
\hat{\sigma} := \frac{\sigma}{p} + \frac{\sigma_1}{q} = \sigma - \frac{\sigma - \sigma_1}{q}.
$$

We can rewrite the reverse of (2.2) as follows:

$$
\int_0^\infty \int_0^\infty h(xy) f(x) g(y) dx \, dy
$$

$$
> k^{\frac{1}{p}}(\sigma) k^{\frac{1}{q}}(\sigma_1) \left[\int_0^\infty x^{p(1 - \hat{\sigma}) - 1} f^p(x) dx \right]^{\frac{1}{p}}
$$

$$
\times \left[\int_0^\infty y^{q(1 - \hat{\sigma}) - 1} g^q(y) dy \right]^{\frac{1}{q}}. \tag{2.60}
$$

By the reverse Hölder inequality with weight (cf. Ref. [120]), we have the following reverse inequality:

$$k(\widehat{\sigma}) = \int_0^\infty h(u)(u^{\frac{\sigma-1}{p}})(u^{\frac{\sigma_1-1}{q}})du$$

$$\geq \left(\int_0^\infty h(u)u^{\sigma-1}du \right)^{\frac{1}{p}} \left(\int_0^\infty h(u)u^{\sigma_1-1}du \right)^{\frac{1}{q}}$$

$$= k^{\frac{1}{p}}(\sigma)k^{\frac{1}{q}}(\sigma_1) > 0. \tag{2.61}$$

If $\sigma - \sigma_1 \in (q\delta_0, -q\delta_0)$, then $\widehat{\sigma} \in (\sigma - \delta_0, \sigma + \delta_0)$, and by Lemma 2.1,

$$k(\widehat{\sigma}) = k\left(\sigma - \frac{\sigma - \sigma_1}{q}\right) \leq k(\sigma - \delta_0) + k(\sigma + \delta_0) < \infty.$$

Lemma 2.22. *For $0 < p < 1$, if there exists a constant $\delta_0 > 0$ such that $k(\sigma \pm \delta_0) < \infty$ and the constant factor $k^{\frac{1}{p}}(\sigma)k^{\frac{1}{q}}(\sigma_1)$ in (2.60) (or the reverse of (2.2)) is the best possible, then for $\sigma - \sigma_1 \in (q\delta_0, -q\delta_0)$, we have $\sigma_1 = \sigma$.*

Proof. If the constant factor $k^{\frac{1}{p}}(\sigma)k^{\frac{1}{q}}(\sigma_1)$ in (2.60) is the best possible, then in view of the reverse of (2.7) (for $\sigma = \widehat{\sigma}$), we have

$$k^{\frac{1}{p}}(\sigma)k^{\frac{1}{q}}(\sigma_1) \geq k(\widehat{\sigma}) \; (\in \mathbf{R}_+),$$

namely, (2.61) keeps the form of equality.

We observe that (2.61) keeps the form of equality if and only if there exist constants A and B such that they are not both zero and (cf. Ref. [120])

$$Au^{\sigma-1} = Bu^{\sigma_1-1} \quad \text{a.e. in } \mathbf{R}_+.$$

Assuming that $A \neq 0$, we have $u^{\sigma-\sigma_1} = \frac{B}{A}$ a.e. in $\mathbf{R}_+$, thus $\sigma - \sigma_1 = 0$, namely, $\sigma_1 = \sigma$.

This completes the proof of the lemma. □

Theorem 2.23. *For $0 < p < 1$, the reverse of inequality (2.2) is equivalent to the following inequalities:*

$$J := \left[\int_0^\infty y^{p(\frac{\sigma}{p} + \frac{\sigma_1}{q}) - 1} \left(\int_0^\infty h(xy) f(x) dx \right)^p dy \right]^{\frac{1}{p}}$$

$$> k^{\frac{1}{p}}(\sigma) k^{\frac{1}{q}}(\sigma_1) \left\{ \int_0^\infty x^{p[1 - (\frac{\sigma}{p} + \frac{\sigma_1}{q})] - 1} f^p(x) dx \right\}^{\frac{1}{p}}, \qquad (2.62)$$

$$J_1 := \left[\int_0^\infty x^{q(\frac{\sigma}{p} + \frac{\sigma_1}{q}) - 1} \left(\int_0^\infty h(xy) g(y) dy \right)^q dx \right]^{\frac{1}{q}}$$

$$> k^{\frac{1}{p}}(\sigma) k^{\frac{1}{q}}(\sigma_1) \left\{ \int_0^\infty y^{q[1 - (\frac{\sigma}{p} + \frac{\sigma_1}{q})] - 1} g^q(y) dy \right\}^{\frac{1}{q}}. \qquad (2.63)$$

The constant factor $k^{\frac{1}{p}}(\sigma) k^{\frac{1}{q}}(\sigma_1)$ in (2.62) and (2.63) is the best possible if and only if the same constant factor in the reverse of (2.2) is the best possible.

In particular, for $\sigma_1 = \sigma$, if there exists a constant $\delta_0 > 0$ such that $k(\sigma \pm \delta_0) < \infty$, then we have the following reverse inequalities equivalent to the reverse of (2.7) with the same best possible constant factor $k(\sigma)$:

$$\left[\int_0^\infty y^{p\sigma - 1} \left(\int_0^\infty h(xy) f(x) dx \right)^p dy \right]^{\frac{1}{p}}$$

$$> k(\sigma) \left[\int_0^\infty x^{p(1-\sigma) - 1} f^p(x) dx \right]^{\frac{1}{p}}, \qquad (2.64)$$

$$\left[\int_0^\infty x^{q\sigma - 1} \left(\int_0^\infty h(xy) g(y) dy \right)^q dx \right]^{\frac{1}{q}}$$

$$> k(\sigma) \left[\int_0^\infty y^{q(1-\sigma) - 1} g^q(y) dy \right]^{\frac{1}{q}}. \qquad (2.65)$$

Proof. If (2.62) is valid, then by the reverse Hölder inequality (cf. Ref. [120]), we have

$$I = \int_0^\infty \left[y^{\frac{-1}{p} + (\frac{\sigma}{p} + \frac{\sigma_1}{q})} \int_0^\infty h(xy) f(x) dx \right] [y^{\frac{1}{p} - (\frac{\sigma}{p} + \frac{\sigma_1}{q})} g(y)] dy$$

$$\geq J \left\{ \int_0^\infty y^{q[1 - (\frac{\sigma}{p} + \frac{\sigma_1}{q})] - 1} g^q(y) dy \right\}^{\frac{1}{q}}. \qquad (2.66)$$

Using (2.62), obtain the reverse of (2.2).

On the other hand, assuming that the reverse of (2.2) is valid, we set

$$g(y) := y^{p(\frac{\sigma}{p} + \frac{\sigma_1}{q}) - 1} \left(\int_0^\infty h(xy) f(x) dx \right)^{p-1} \quad (y > 0).$$

Then, it follows that

$$J^p = \int_0^\infty y^{q[1 - (\frac{\sigma}{p} + \frac{\sigma_1}{q})] - 1} g^q(y) dy = I. \tag{2.67}$$

If $J = \infty$, then (2.62) is naturally valid; if $J = 0$, then it is impossible to make (2.62) valid, namely, $J > 0$. Suppose that $0 < J < \infty$. By the reverse of (2.2), we have

$$J^p = \int_0^\infty y^{q[1 - (\frac{\sigma}{p} + \frac{\sigma_1}{q})] - 1} g^q(y) dy = I$$

$$> k^{\frac{1}{p}}(\sigma) k^{\frac{1}{q}}(\sigma_1) \left\{ \int_0^\infty x^{p[1 - (\frac{\sigma}{p} + \frac{\sigma_1}{q})] - 1} f^p(x) dx \right\}^{\frac{1}{p}} J^{p-1},$$

$$J = \left\{ \int_0^\infty y^{q[1 - (\frac{\sigma}{p} + \frac{\sigma_1}{q})] - 1} g^q(y) dy \right\}^{\frac{1}{p}}$$

$$> k^{\frac{1}{p}}(\sigma) k^{\frac{1}{q}}(\sigma_1) \left\{ \int_0^\infty x^{p[1 - (\frac{\sigma}{p} + \frac{\sigma_1}{q})] - 1} f^p(x) dx \right\}^{\frac{1}{p}},$$

namely, (2.62) follows, which is equivalent to the reverse of (2.2).

Similarly, we can show that (2.63) is equivalent to the reverse of (2.2). Hence, we have the reverse of (2.2): Inequalities (2.62) and (2.63) are equivalent.

If the constant factor $k^{\frac{1}{p}}(\sigma) k^{\frac{1}{q}}(\sigma_1)$ in the reverse of (2.2) is the best possible, then the same constant factor in (2.62) is also the best possible. Otherwise, by (2.66), we would reach a contradiction that the same constant factor in the reverse of (2.2) is not the best possible. If the constant factor $k^{\frac{1}{p}}(\sigma) k^{\frac{1}{q}}(\sigma_1)$ in (2.62) is the best possible, then the same constant factor in the reverse of (2.2) is also the best possible. Otherwise, by (2.67), we would reach a contradiction that the same constant factor in (2.62) is not the best possible.

Similarly, we can show that the constant factor $k^{\frac{1}{p}}(\sigma)k^{\frac{1}{q}}(\sigma_1)$ in the reverse of (2.2) is the best possible if and only if the same constant factor in (2.63) is the best possible. Therefore, the constant factor $k^{\frac{1}{p}}(\sigma)k^{\frac{1}{q}}(\sigma_1)$ in (2.62) and (2.63) is the best possible if and only if the same constant factor in the reverse of (2.2) is the best possible.

This completes the proof of the theorem.   □

Theorem 2.24. *For $0 < p < 1$, if there exists a constant $\delta_0 > 0$ such that $k(\sigma \pm \delta_0) < \infty$, then the following statements, (i), (ii), (iii), and (iv), are equivalent:*

(i) both

$$k^{\frac{1}{p}}(\sigma)k^{\frac{1}{q}}(\sigma_1) \quad \text{and} \quad k\left(\frac{\sigma}{p} + \frac{\sigma_1}{q}\right)$$

are independent of p and q;

(ii) $k^{\frac{1}{p}}(\sigma)k^{\frac{1}{q}}(\sigma_1) = k(\frac{\sigma}{p} + \frac{\sigma_1}{q})$;

(iii) for $\sigma - \sigma_1 \in (q\delta_0, -q\delta_0)$, we have $\sigma_1 = \sigma$;

(iv) the constant factor $k^{\frac{1}{p}}(\sigma)k^{\frac{1}{q}}(\sigma_1)$ in the reverse of (2.2), which is the inequalities (2.62) and (2.63), is the best possible.

Proof.   (i) $\Rightarrow$ (ii): We have

$$k^{\frac{1}{p}}(\sigma)k^{\frac{1}{q}}(\sigma_1) = \lim_{p \to 1^-} \lim_{q \to -\infty} k^{\frac{1}{p}}(\sigma)k^{\frac{1}{q}}(\sigma_1) = k(\sigma).$$

By Lemma 2.1, we derive that

$$k\left(\frac{\sigma}{p} + \frac{\sigma_1}{q}\right) = \lim_{p \to 1^-} \lim_{q \to -\infty} k\left(\sigma + \frac{\sigma_1 - \sigma}{q}\right)$$

$$= k(\sigma) = k^{\frac{1}{p}}(\sigma)k^{\frac{1}{q}}(\sigma_1).$$

(ii) $\Rightarrow$ (iii): If

$$k^{\frac{1}{p}}(\sigma)k^{\frac{1}{q}}(\sigma_1) = k\left(\frac{\sigma}{p} + \frac{\sigma_1}{q}\right),$$

then (2.61) keeps the form of equality. In view of the proof of Lemma 2.22, for

$$\sigma - \sigma_1 \in (q\delta_0, -q\delta_0),$$

we have $\sigma_1 = \sigma$.

(iii) $\Rightarrow$ (i): For $\sigma_1 = \sigma$, both $k^{\frac{1}{p}}(\sigma)k^{\frac{1}{q}}(\sigma_1)$ and $k(\frac{\sigma}{p} + \frac{\sigma_1}{q})$ are independent of p and q, which are equal to $k(\sigma)$.

Hence, we have (i) $\Leftrightarrow$ (ii) $\Leftrightarrow$ (iii).

(iii) $\Rightarrow$ (iv): For $\sigma_1 = \sigma$, by Theorem 2.23, the constant factor $k^{\frac{1}{p}}(\sigma)k^{\frac{1}{q}}(\sigma_1)(= k(\sigma))$ is the best possible in the reverse of (2.2), that is, in inequalities (2.62) and (2.63).

(iv) $\Rightarrow$ (iii): Since $\sigma - \sigma_1 \in (q\delta_0, -q\delta_0)$, by Lemma 2.22, we have $\sigma_1 = \sigma$.

Hence, we have (iii) $\Leftrightarrow$ (iv).

Therefore, the statements (i), (ii), (iii), and (iv) are equivalent.

This completes the proof of the theorem. $\qquad\square$

If $k_\lambda(x, y)(\geq 0)$ is a homogeneous function of degree $-\lambda$ satisfying

$$k_\lambda(ux, uy) = u^{-\lambda}k_\lambda(x, y) \ (u, x, y > 0),$$

then setting $h(u) = k_\lambda(1, u)$ and replacing x by $\frac{1}{x}$ and $x^{\lambda-2}f(\frac{1}{x})$ by $f(x)$ in Theorems 2.23 and 2.24, for $\sigma_1 = \lambda - \mu$, we have the following.

Corollary 2.25. *For $0 < p < 1$, if*

$$k_\lambda(\gamma) := \int_0^\infty k_\lambda(1, u)u^{\sigma-1}du \in \mathbf{R}_+ \ (\gamma = \sigma, \lambda - \mu),$$

$$0 < \int_0^\infty x^{p[1-(\frac{\lambda-\sigma}{p}+\frac{\mu}{q})]-1}f^p(x)dx < \infty, \ and$$

$$0 < \int_0^\infty y^{q[1-(\frac{\sigma}{p}+\frac{\lambda-\mu}{q})]-1}g^q(y)dy < \infty,$$

then we have the following reverse equivalent Hilbert-type integral inequalities with a homogeneous kernel:

$$\int_0^\infty \int_0^\infty k_\lambda(x, y)f(x)g(y)dx\,dy$$

$$> k_\lambda^{\frac{1}{p}}(\sigma)k_\lambda^{\frac{1}{q}}(\lambda - \mu)\left\{\int_0^\infty x^{p[1-(\frac{\lambda-\sigma}{p}+\frac{\mu}{q})]-1}f^p(x)dx\right\}^{\frac{1}{p}}$$

$$\times \left\{\int_0^\infty y^{q[1-(\frac{\sigma}{p}+\frac{\lambda-\mu}{q})]-1}g^q(y)dy\right\}^{\frac{1}{q}}, \tag{2.68}$$

$$\left[\int_0^\infty y^{p(\frac{\sigma}{p}+\frac{\lambda-\mu}{q})-1}\left(\int_0^\infty k_\lambda(x,y)f(x)dx\right)^p dy\right]^{\frac{1}{p}}$$

$$> k_\lambda^{\frac{1}{p}}(\sigma)k_\lambda^{\frac{1}{q}}(\lambda-\mu)\left\{\int_0^\infty x^{p[1-(\frac{\lambda-\sigma}{p}+\frac{\mu}{q})]-1}f^p(x)dx\right\}^{\frac{1}{p}}, \quad (2.69)$$

$$\left[\int_0^\infty x^{q(\frac{\lambda-\sigma}{p}+\frac{\mu}{q})-1}\left(\int_0^\infty k_\lambda(x,y)g(y)dy\right)^q dx\right]^{\frac{1}{q}}$$

$$> k_\lambda^{\frac{1}{p}}(\sigma)k_\lambda^{\frac{1}{q}}(\lambda-\mu)\left\{\int_0^\infty y^{q[1-(\frac{\sigma}{p}+\frac{\lambda-\mu}{q})]-1}g^q(y)dy\right\}^{\frac{1}{q}}. \quad (2.70)$$

Moreover, the constant factor

$$k_\lambda^{\frac{1}{p}}(\sigma)k_\lambda^{\frac{1}{q}}(\lambda-\mu)$$

is the best possible in (2.68) if and only if the same constant factor in (2.69) and (2.70) is the best possible.

In particular, for $\mu+\sigma=\lambda$, if there exists a constant $\delta_0>0$ such that $k_\lambda(\sigma\pm\delta_0)<\infty$, then we have the following equivalent Hilbert-type integral inequalities with the best possible constant factor $k_\lambda(\sigma)$:

$$\int_0^\infty\int_0^\infty k_\lambda(x,y)f(x)g(y)dx\,dy$$

$$> k_\lambda(\sigma)\left[\int_0^\infty x^{p(1-\mu)-1}f^p(x)dx\right]^{\frac{1}{p}}$$

$$\times\left[\int_0^\infty y^{q(1-\sigma)-1}g^q(y)dy\right]^{\frac{1}{q}}, \quad (2.71)$$

$$\left[\int_0^\infty y^{p\sigma-1}\left(\int_0^\infty k_\lambda(x,y)f(x)dx\right)^p dy\right]^{\frac{1}{p}}$$

$$> k_\lambda(\sigma)\left[\int_0^\infty x^{p(1-\mu)-1}f^p(x)dx\right]^{\frac{1}{p}}, \quad (2.72)$$

$$\left[\int_0^\infty x^{q\mu-1}\left(\int_0^\infty k_\lambda(x,y)g(y)dy\right)^q dx\right]^{\frac{1}{q}}$$

$$> k_\lambda(\sigma)\left[\int_0^\infty y^{q(1-\sigma)-1}g^q(y)dy\right]^{\frac{1}{q}}. \tag{2.73}$$

Corollary 2.26. *For $0 < p < 1$, if there exists a constant $\delta_0 > 0$ such that $k_\lambda(\sigma \pm \delta_0) < \infty$, then the following statements, (I), (II), (III), and (IV), are equivalent:*

(I) both

$$k_\lambda^{\frac{1}{p}}(\sigma)k_\lambda^{\frac{1}{q}}(\lambda - \mu)$$

and

$$k_\lambda\left(\frac{\sigma}{p} + \frac{\lambda - \mu}{q}\right)$$

are independent of p and q;

(II) $k_\lambda^{\frac{1}{p}}(\sigma)k_\lambda^{\frac{1}{q}}(\lambda - \mu) = k_\lambda(\frac{\sigma}{p} + \frac{\lambda-\mu}{q})$;

(III) for $\mu + \sigma - \lambda \in (q\delta_0, -q\delta_0)$, we have $\mu + \sigma = \lambda$;

(IV) the constant factor

$$k_\lambda^{\frac{1}{p}}(\sigma)k_\lambda^{\frac{1}{q}}(\lambda - \mu)$$

in (2.68)–(2.70) is the best possible.

Replacing x (resp. y) by x^α (resp. y^β), then replacing $x^{\alpha-1}f(x^\alpha)$ (resp. $y^{\beta-1}g(y^\beta)$) by $f(x)$ (resp. $g(y)$) in Corollary 2.26 and Theorem 2.23, by simplification, we deduce the following corollary.

Corollary 2.27. *For $0 < p < 1$, if $k_\lambda(\gamma) \in \mathbf{R}_+$ $(\gamma = \sigma, \lambda - \mu)$,*

$$0 < \int_0^\infty x^{p[1-\alpha(\frac{\lambda-\sigma}{p}+\frac{\mu}{q})]-1}f^p(x)dx < \infty, \quad and$$

$$0 < \int_0^\infty y^{q[1-\beta(\frac{\sigma}{p}+\frac{\lambda-\mu}{q})]-1}g^q(y)dy < \infty,$$

then we have the following reverse equivalent integral inequalities with a homogeneous kernel:

$$\int_0^\infty \int_0^\infty k_\lambda(x^\alpha, y^\beta) f(x) g(y) dx\, dy$$

$$> \frac{1}{\alpha^{1/q} \beta^{1/p}} k_\lambda^{\frac{1}{p}}(\sigma) k_\lambda^{\frac{1}{q}}(\lambda - \mu) \left\{ \int_0^\infty x^{p[1 - \alpha(\frac{\lambda - \sigma}{p} + \frac{\mu}{q})] - 1} f^p(x) dx \right\}^{\frac{1}{p}}$$

$$\times \left\{ \int_0^\infty y^{q[1 - \beta(\frac{\sigma}{p} + \frac{\lambda - \mu}{q})] - 1} g^q(y) dy \right\}^{\frac{1}{q}}, \tag{2.74}$$

$$\left[\int_0^\infty y^{p\beta(\frac{\sigma}{p} + \frac{\lambda - \mu}{q}) - 1} \left(\int_0^\infty k_\lambda(x^\alpha, y^\beta) f(x) dx \right)^p dy \right]^{\frac{1}{p}}$$

$$> \frac{1}{\alpha^{1/q} \beta^{1/p}} k_\lambda^{\frac{1}{p}}(\sigma) k_\lambda^{\frac{1}{q}}(\lambda - \mu)$$

$$\times \left\{ \int_0^\infty x^{p[1 - \alpha(\frac{\lambda - \sigma}{p} + \frac{\mu}{q})] - 1} f^p(x) dx \right\}^{\frac{1}{p}}, \tag{2.75}$$

$$\left[\int_0^\infty x^{q\alpha(\frac{\lambda - \sigma}{p} + \frac{\mu}{q}) - 1} \left(\int_0^\infty k_\lambda(x^\alpha, y^\beta) g(y) dy \right)^q dx \right]^{\frac{1}{q}}$$

$$> \frac{1}{\alpha^{1/q} \beta^{1/p}} k_\lambda^{\frac{1}{p}}(\sigma) k_\lambda^{\frac{1}{q}}(\lambda - \mu)$$

$$\times \left\{ \int_0^\infty y^{q[1 - \beta(\frac{\sigma}{p} + \frac{\lambda - \mu}{q})] - 1} g^q(y) dy \right\}^{\frac{1}{q}}. \tag{2.76}$$

Moreover, the constant factor

$$\frac{1}{\alpha^{1/q} \beta^{1/p}} k_\lambda^{\frac{1}{p}}(\sigma) k_\lambda^{\frac{1}{q}}(\lambda - \mu)$$

is the best possible in (2.74) *if and only if the same constant factor in* (2.75) *and* (2.76) *is the best possible.*

In particular, for $\mu + \sigma = \lambda$, if there exists a constant $\delta_0 > 0$ such that $k_\lambda(\sigma \pm \delta_0) < \infty$, then we have the following equivalent inequalities with the best possible constant factor

$$\frac{k_\lambda(\sigma)}{\alpha^{1/q}\beta^{1/p}} : \int_0^\infty \int_0^\infty k_\lambda(x^\alpha, y^\beta) f(x)g(y)dx\,dy$$

$$> \frac{k_\lambda(\sigma)}{\alpha^{1/q}\beta^{1/p}} \left[\int_0^\infty x^{p(1-\alpha\mu)-1} f^p(x)dx \right]^{\frac{1}{p}}$$

$$\times \left[\int_0^\infty y^{q(1-\beta\sigma)-1} g^q(y)dy \right]^{\frac{1}{q}}, \tag{2.77}$$

$$\left[\int_0^\infty y^{p\beta\sigma-1} \left(\int_0^\infty k_\lambda(x^\alpha, y^\beta) f(x)dx \right)^p dy \right]^{\frac{1}{p}}$$

$$> \frac{k_\lambda(\sigma)}{\alpha^{1/q}\beta^{1/p}} \left[\int_0^\infty x^{p(1-\alpha\mu)-1} f^p(x)dx \right]^{\frac{1}{p}}, \tag{2.78}$$

$$\left[\int_0^\infty x^{q\alpha\mu-1} \left(\int_0^\infty k_\lambda(x^\alpha, y^\beta) g(y)dy \right)^q dx \right]^{\frac{1}{q}}$$

$$> \frac{k_\lambda(\sigma)}{\alpha^{1/q}\beta^{1/p}} \left[\int_0^\infty y^{q(1-\beta\sigma)-1} g^q(y)dy \right]^{\frac{1}{q}}. \tag{2.79}$$

Corollary 2.28. *For* $0 < p < 1$, *if* $k(\gamma) \in \mathbf{R}_+$ $(\gamma = \sigma, \sigma_1)$,

$$0 < \int_0^\infty x^{p[1-\alpha(\frac{\sigma}{p}+\frac{\sigma_1}{q})]-1} f^p(x)dx < \infty, \ and$$

$$0 < \int_0^\infty y^{q[1-\beta(\frac{\sigma}{p}+\frac{\sigma_1}{q})]-1} g^q(y)dy < \infty,$$

then we have the following reverse equivalent Hilbert-type integral inequalities with a nonhomogeneous kernel:

$$\int_0^\infty \int_0^\infty h(x^\alpha y^\beta) f(x)g(y)dx\,dy$$

$$> \frac{1}{\alpha^{1/q}\beta^{1/p}} k^{\frac{1}{p}}(\sigma) k^{\frac{1}{q}}(\sigma_1) \left\{ \int_0^\infty x^{p[1-\alpha(\frac{\sigma}{p}+\frac{\sigma_1}{q})]-1} f^p(x)dx \right\}^{\frac{1}{p}}$$

$$\times \left\{ \int_0^\infty y^{q[1-\beta(\frac{\sigma}{p}+\frac{\sigma_1}{q})]-1} g^q(y)dy \right\}^{\frac{1}{q}}, \tag{2.80}$$

$$\left[\int_0^\infty y^{p\beta(\frac{\sigma}{p}+\frac{\sigma_1}{q})-1}\left(\int_0^\infty h(x^\alpha y^\beta)f(x)dx\right)^p dy\right]^{\frac{1}{p}}$$

$$> \frac{1}{\alpha^{1/q}\beta^{1/p}}k^{\frac{1}{p}}(\sigma)k^{\frac{1}{q}}(\sigma_1)$$

$$\times\left\{\int_0^\infty x^{p[1-\alpha(\frac{\sigma}{p}+\frac{\sigma_1}{q})]-1}f^p(x)dx\right\}^{\frac{1}{p}}, \tag{2.81}$$

$$\left[\int_0^\infty x^{q\alpha(\frac{\sigma}{p}+\frac{\sigma_1}{q})-1}\left(\int_0^\infty h(x^\alpha y^\beta)g(y)dy\right)^q dx\right]^{\frac{1}{q}}$$

$$> \frac{1}{\alpha^{1/q}\beta^{1/p}}k^{\frac{1}{p}}(\sigma)k^{\frac{1}{q}}(\sigma_1)\left\{\int_0^\infty y^{q[1-\beta(\frac{\sigma}{p}+\frac{\sigma_1}{q})]-1}g^q(y)dy\right\}^{\frac{1}{q}}.$$

$$\tag{2.82}$$

Moreover, the constant factor

$$\frac{1}{\alpha^{1/q}\beta^{1/p}}k^{\frac{1}{p}}(\sigma)k^{\frac{1}{q}}(\sigma_1)$$

in (2.80) *is the best possible if and only if the same constant factor in* (2.81) *and* (2.82) *is the best possible.*

In particular, for $\sigma_1 = \sigma$, if there exists a constant $\delta_0 > 0$ such that $k(\sigma \pm \delta_0) < \infty$, then we have the following reverse equivalent Hilbert-type integral inequalities with a nonhomogeneous kernel and the best possible constant factor

$$\frac{k(\sigma)}{\alpha^{1/q}\beta^{1/p}} : \int_0^\infty \int_0^\infty h(x^\alpha y^\beta)f(x)g(y)dx\,dy$$

$$> \frac{k(\sigma)}{\alpha^{1/q}\beta^{1/p}}\left[\int_0^\infty x^{p(1-\alpha\sigma)-1}f^p(x)dx\right]^{\frac{1}{p}}$$

$$\times\left[\int_0^\infty y^{q(1-\beta\sigma)-1}g^q(y)dy\right]^{\frac{1}{q}}, \tag{2.83}$$

$$\left[\int_0^\infty y^{p\beta\sigma-1}\left(\int_0^\infty h(x^\alpha y^\beta)f(x)dx\right)^p dy\right]^{\frac{1}{p}}$$

$$> \frac{k(\sigma)}{\alpha^{1/q}\beta^{1/p}}\left[\int_0^\infty x^{p(1-\alpha\sigma)-1}f^p(x)dx\right]^{\frac{1}{p}}, \tag{2.84}$$

$$\left[\int_0^\infty x^{q\alpha\sigma-1}\left(\int_0^\infty h(x^\alpha y^\beta)g(y)dy\right)^q dx\right]^{\frac{1}{q}}$$

$$> \frac{k(\sigma)}{\alpha^{1/q}\beta^{1/p}}\left[\int_0^\infty y^{q(1-\beta\sigma)-1}g^q(y)dy\right]^{\frac{1}{q}}. \tag{2.85}$$

Example 2.29. (i) Setting

$$h(u) = k_\lambda(1, u) = \frac{1}{(1+u)^\lambda} \quad (\lambda > 0),$$

we obtain

$$k(\gamma) = k_\lambda(\gamma) = \int_0^\infty \frac{u^{\gamma-1}du}{(1+u)^\lambda} = B(\gamma, \lambda - \gamma) \quad (0 < \gamma = \mu, \sigma < \lambda).$$

By (2.77)–(2.79), we deduce the following reverse equivalent extended Hardy–Hilbert integral inequalities with a homogeneous kernel and the best possible constant factor

$$\frac{B(\mu,\sigma)}{\alpha^{1/q}\beta^{1/p}} : \int_0^\infty \int_0^\infty \frac{1}{(x^\alpha + y^\beta)^\lambda}f(x)g(y)dx\,dy$$

$$> \frac{B(\mu,\sigma)}{\alpha^{1/q}\beta^{1/p}}\left[\int_0^\infty x^{p(1-\alpha\mu)-1}f^p(x)dx\right]^{\frac{1}{p}}$$

$$\times \left[\int_0^\infty y^{q(1-\beta\sigma)-1}g^q(y)dy\right]^{\frac{1}{q}}, \tag{2.86}$$

$$\left\{ \int_0^\infty y^{p\beta\sigma-1} \left[\int_0^\infty \frac{1}{(x^\alpha+y^\beta)^\lambda} f(x)dx \right]^p dy \right\}^{\frac{1}{p}}$$

$$> \frac{B(\mu,\sigma)}{\alpha^{1/q}\beta^{1/p}} \left[\int_0^\infty x^{p(1-\alpha\mu)-1} f^p(x)dx \right]^{\frac{1}{p}}, \qquad (2.87)$$

$$\left\{ \int_0^\infty x^{q\alpha\mu-1} \left[\int_0^\infty \frac{1}{(x^\alpha+y^\beta)^\lambda} g(y)dy \right]^q dx \right\}^{\frac{1}{q}}$$

$$> \frac{B(\mu,\sigma)}{\alpha^{1/q}\beta^{1/p}} \left[\int_0^\infty y^{q(1-\beta\sigma)-1} g^q(y)dy \right]^{\frac{1}{q}}. \qquad (2.88)$$

In particular, for $\alpha = \beta = 1$, we have

$$\int_0^\infty \int_0^\infty \frac{1}{(x+y)^\lambda} f(x)g(y)dx\,dy$$

$$> B(\mu,\sigma) \left[\int_0^\infty x^{p(1-\mu)-1} f^p(x)dx \right]^{\frac{1}{p}}$$

$$\times \left[\int_0^\infty y^{q(1-\sigma)-1} g^q(y)dy \right]^{\frac{1}{q}}, \qquad (2.89)$$

$$\left\{ \int_0^\infty y^{p\sigma-1} \left[\int_0^\infty \frac{1}{(x+y)^\lambda} f(x)dx \right]^p dy \right\}^{\frac{1}{p}}$$

$$> B(\mu,\sigma) \left[\int_0^\infty x^{p(1-\mu)-1} f^p(x)dx \right]^{\frac{1}{p}}, \qquad (2.90)$$

$$\left\{ \int_0^\infty x^{q\mu-1} \left[\int_0^\infty \frac{1}{(x+y)^\lambda} g(y)dy \right]^q dx \right\}^{\frac{1}{q}}$$

$$> B(\mu,\sigma) \left[\int_0^\infty y^{q(1-\sigma)-1} g^q(y)dy \right]^{\frac{1}{q}}. \qquad (2.91)$$

By (2.83)–(2.85), we have the following reverse equivalent extended Hardy–Hilbert integral inequalities with a nonhomogeneous kernel and the best possible constant factor

$$\frac{B(\lambda - \sigma, \sigma)}{\alpha^{1/q}\beta^{1/p}} : \int_0^\infty \int_0^\infty \frac{1}{(1 + x^\alpha y^\beta)^\lambda} f(x)g(y)dx\,dy$$

$$> \frac{B(\lambda - \sigma, \sigma)}{\alpha^{1/q}\beta^{1/p}} \left[\int_0^\infty x^{p(1-\alpha\sigma)-1} f^p(x)dx \right]^{\frac{1}{p}}$$

$$\times \left[\int_0^\infty y^{q(1-\beta\sigma)-1} g^q(y)dy \right]^{\frac{1}{q}}, \tag{2.92}$$

$$\left\{ \int_0^\infty y^{p\beta\sigma-1} \left[\int_0^\infty \frac{1}{(1 + x^\alpha y^\beta)^\lambda} f(x)dx \right]^p dy \right\}^{\frac{1}{p}}$$

$$> \frac{B(\lambda - \sigma, \sigma)}{\alpha^{1/q}\beta^{1/p}} \left[\int_0^\infty x^{p(1-\alpha\sigma)-1} f^p(x)dx \right]^{\frac{1}{p}}, \tag{2.93}$$

$$\left\{ \int_0^\infty x^{q\alpha\sigma-1} \left[\int_0^\infty \frac{1}{(1 + x^\alpha y^\beta)^\lambda} g(y)dy \right]^q dx \right\}^{\frac{1}{q}}$$

$$> \frac{B(\lambda - \sigma, \sigma)}{\alpha^{1/q}\beta^{1/p}} \left[\int_0^\infty y^{q(1-\beta\sigma)-1} g^q(y)dy \right]^{\frac{1}{q}}. \tag{2.94}$$

In particular, for $\alpha = \beta = 1$, we have

$$\int_0^\infty \int_0^\infty \frac{1}{(1 + xy)^\lambda} f(x)g(y)dx\,dy$$

$$> B(\lambda - \sigma, \sigma) \left[\int_0^\infty x^{p(1-\sigma)-1} f^p(x)dx \right]^{\frac{1}{p}}$$

$$\times \left[\int_0^\infty y^{q(1-\sigma)-1} g^q(y)dy \right]^{\frac{1}{q}}, \tag{2.95}$$

$$\left\{\int_0^\infty y^{p\sigma-1}\left[\int_0^\infty \frac{1}{(1+xy)^\lambda}f(x)dx\right]^p dy\right\}^{\frac{1}{p}}$$

$$> B(\lambda-\sigma,\sigma)\left[\int_0^\infty x^{p(1-\sigma)-1}f^p(x)dx\right]^{\frac{1}{p}}, \qquad (2.96)$$

$$\left\{\int_0^\infty x^{q\sigma-1}\left[\int_0^\infty \frac{1}{(1+xy)^\lambda}g(y)dy\right]^q dx\right\}^{\frac{1}{q}}$$

$$> B(\lambda-\sigma,\sigma)\left[\int_0^\infty y^{q(1-\beta\sigma)-1}g^q(y)dy\right]^{\frac{1}{q}}. \qquad (2.97)$$

Chapter 3

A New Hilbert-Type Integral Inequality Involving One Derivative Function

In this chapter, using weight functions, the idea of introducing parameters and techniques of real analysis, and applying the extended Hardy–Hilbert integral inequality, we present a new Hilbert-type integral inequality with a homogeneous kernel involving one derivative function and the beta function. The equivalent statements of the best possible constant factor related to several parameters are provided. The equivalent form, the case of a nonhomogeneous kernel, a few particular inequalities, the operator expressions, and the reverses are considered.

3.1 Some Lemmas

Hereinafter in this chapter, we assume that $p > 0$ $(p \neq 1)$, $\frac{1}{p} + \frac{1}{q} = 1$, $\lambda > 0, \lambda_i \in (0, \lambda)$ $(i = 1, 2)$, $a := \lambda - \lambda_1 - \lambda_2$, $f(x)$ is a nonnegative measurable function in $\mathbf{R}_+ = (0, \infty)$, and $g(y)$ is a nonnegative increasing function in $\mathbf{R}_+$, with

$$g'(x) \geq 0, \ g(y) = o(1) \ (y \to 0^+)$$

and

$$e^{-ty}g(y) \in L(\mathbf{R}_+) = \{f; \ f \text{ is an } L\text{-integrable function in } \mathbf{R}_+\}$$

satisfying

$$0 < \int_0^\infty x^{p(1-\lambda_1)-a-1} f^p(x)dx < \infty$$

and

$$0 < \int_0^\infty y^{q(1-\lambda_2)-a-1} g'^q(y)dy < \infty.$$

By the definition of the gamma function,

$$\Gamma(\alpha) := \int_0^\infty e^{-u} u^{\alpha-1} du \ (\alpha > 0)$$

for $\lambda, x, y > 0$, the following expression holds (cf. Ref. [123]):

$$\frac{1}{(x+y)^\lambda} = \frac{1}{\Gamma(\lambda)} \int_0^\infty e^{-(x+y)t} t^{\lambda-1} dt. \tag{3.1}$$

We still have

$$\Gamma(\alpha+1) = \alpha\Gamma(\alpha) \ (\alpha > 0).$$

Lemma 3.1. *For $t > 0$, we have the following inequality:*

$$\int_0^\infty e^{-ty} g(y)dy = \frac{1}{t} \int_0^\infty e^{-ty} g'(y)dy. \tag{3.2}$$

Proof. Since $g(y) = o(1) \ (y \to 0^+)$, we obtain that

$$\int_0^\infty e^{-ty} g'(y)dy = \int_0^\infty e^{-ty} dg(y)$$

$$= e^{-ty} g(y)|_0^\infty - \int_0^\infty g(y) de^{-ty}$$

$$= \lim_{y\to\infty} e^{-ty} g(y) + t \int_0^\infty e^{-ty} g(y)dy.$$

In view of $e^{-ty} g(y) \in L(\mathbf{R}_+)$, we deduce that

$$e^{-ty} g(y) \to 0 \ (y \to \infty),$$

and thus,

$$t \int_0^\infty e^{-ty} g(y) dy = \int_0^\infty e^{-ty} g'(y) dy,$$

namely, expression (3.2) follows.

This completes the proof of the lemma. □

Lemma 3.2. *Define the following weight functions:*

$$\varpi_\lambda(\lambda_2, x) := x^{\lambda - \lambda_2} \int_0^\infty \frac{t^{\lambda_2 - 1}}{(x + t)^\lambda} dt \ (x \in \mathbf{R}_+), \tag{3.3}$$

$$\omega_\lambda(\lambda_1, y) := y^{\lambda - \lambda_1} \int_0^\infty \frac{t^{\lambda_1 - 1}}{(t + y)^\lambda} dt \ (y \in \mathbf{R}_+). \tag{3.4}$$

We have the following expressions:

$$\varpi_\lambda(\lambda_2, x) = B(\lambda_2, \lambda - \lambda_2) \ (x \in \mathbf{R}_+), \tag{3.5}$$

$$\omega_\lambda(\lambda_1, y) = B(\lambda_1, \lambda - \lambda_1) \ (y \in \mathbf{R}_+), \tag{3.6}$$

where

$$B(u, v) = \int_0^\infty \frac{t^{u-1}}{(1 + t)^{u+v}} dt \ (u, v > 0)$$

is the beta function satisfying

$$B(u, v) = \frac{1}{\Gamma(u + v)} \Gamma(u) \Gamma(v).$$

Proof. Setting $u = \frac{t}{x}$, we obtain that

$$\varpi_\lambda(\lambda_2, x) = x^{\lambda - \lambda_2} \int_0^\infty \frac{(ux)^{\lambda_2 - 1}}{(x + ux)^\lambda} x \, du$$

$$= \int_0^\infty \frac{u^{\lambda_2 - 1}}{(1 + u)^\lambda} du = B(\lambda_2, \lambda - \lambda_2),$$

namely, (3.5) follows. We similarly derive (3.6).

This completes the proof of the lemma. □

Lemma 3.3. *For $p > 1$, we have the following extended Hardy–Hilbert integral inequality:*

$$I_1 := \int_0^\infty \int_0^\infty \frac{f(x)g'(y)}{(x+y)^\lambda} dx\, dy < B^{\frac{1}{p}}(\lambda_2, \lambda - \lambda_2) B^{\frac{1}{q}}(\lambda_1, \lambda - \lambda_1)$$

$$\times \left[\int_0^\infty x^{p(1-\lambda_1)-a-1} f^p(x) dx \right]^{\frac{1}{p}}$$

$$\times \left[\int_0^\infty y^{q(1-\lambda_2)-a-1} g'^q(y) dy \right]^{\frac{1}{q}}. \tag{3.7}$$

Proof. By Hölder's inequality and Fubini's theorem (cf. Ref. [119, 120]), we obtain

$$I_1 = \int_0^\infty \int_0^\infty \frac{1}{(x+y)^\lambda} \left[\frac{y^{(\lambda_2-1)/p}}{x^{(\lambda_1-1)/q}} f(x) \right] \left[\frac{x^{(\lambda_1-1)/q}}{y^{(\lambda_2-1)/p}} g'(y) \right] dx\, dy$$

$$\leq \left\{ \int_0^\infty \int_0^\infty \frac{1}{(x+y)^\lambda} \frac{y^{\lambda_2-1}}{x^{(\lambda_1-1)(p-1)}} f^p(x) dy\, dx \right\}^{\frac{1}{p}}$$

$$\times \left\{ \int_0^\infty \int_0^\infty \frac{1}{(x+y)^\lambda} \frac{x^{\lambda_1-1}}{y^{(\lambda_2-1)(q-1)}} g'^q(y) dx\, dy \right\}^{\frac{1}{q}}$$

$$= \left[\int_0^\infty \varpi_\lambda(\lambda_2, x) x^{p(1-\lambda_1)-a-1} f^p(x) dx \right]^{\frac{1}{p}}$$

$$\times \left[\int_0^\infty \omega_\lambda(\lambda_1, y) y^{q(1-\lambda_2)-a-1} g'^q(y) dy \right]^{\frac{1}{q}}. \tag{3.8}$$

If (3.8) keeps the form of equality, then there exist constants A and B such that they are not both zero, satisfying

$$A \frac{y^{\lambda_2-1}}{x^{(\lambda_1-1)(p-1)}} f^p(x) = B \frac{x^{\lambda_1-1}}{y^{(\lambda_2-1)(q-1)}} g'^q(y) \quad \text{a.e. in } \mathbf{R}_+^2.$$

We assume that $A \neq 0$. Then, there exists a $y \in \mathbf{R}_+$ such that

$$x^{p(1-\lambda_1)-a-1} f^p(x) = \frac{B g'^q(y)}{A y^{q(\lambda_2-1)}} x^{-a-1} \quad \text{a.e. in } \mathbf{R}_+.$$

Since for any $a = \lambda - \lambda_1 - \lambda_2 \in \mathbf{R}$,

$$\int_0^\infty x^{-a-1} dx = \infty,$$

the above expression contradicts the fact that

$$0 < \int_0^\infty x^{p(1-\lambda_1)-a-1} f^p(x) dx < \infty.$$

Therefore, by (3.5) and (3.6), we have (3.7).
This completes the proof of the lemma. □

3.2 Main Results and Equivalent Statements

Theorem 3.4. *For $p > 1$, we have the following Hilbert-type integral inequality involving one derivative function:*

$$I := \int_0^\infty \int_0^\infty \frac{f(x)g(y)}{(x+y)^{\lambda+1}} dx\, dy$$

$$< \frac{1}{\lambda} B^{\frac{1}{p}}(\lambda_2, \lambda - \lambda_2) B^{\frac{1}{q}}(\lambda_1, \lambda - \lambda_1)$$

$$\times \left[\int_0^\infty x^{p(1-\lambda_1)-a-1} f^p(x) dx \right]^{\frac{1}{p}}$$

$$\times \left[\int_0^\infty y^{q(1-\lambda_2)-a-1} g'^q(y) dy \right]^{\frac{1}{q}}. \tag{3.9}$$

In particular, for $\lambda_1 + \lambda_2 = \lambda$ (or $a = 0$), we reduce (3.9) to the following:

$$\int_0^\infty \int_0^\infty \frac{f(x)g(y)}{(x+y)^{\lambda+1}} dx\, dy$$

$$< \frac{1}{\lambda} B(\lambda_1, \lambda_2) \left[\int_0^\infty x^{p(1-\lambda_1)-1} f^p(x) dx \right]^{\frac{1}{p}}$$

$$\times \left[\int_0^\infty y^{q(1-\lambda_2)-1} g'^q(y) dy \right]^{\frac{1}{q}}, \tag{3.10}$$

where the constant factor $\frac{1}{\lambda} B(\lambda_1, \lambda_2)$ is the best possible.

Proof. Using Fubini's theorem (cf. Ref. [119]), (3.1) and (3.2), we obtain that

$$
\begin{aligned}
I &= \frac{1}{\Gamma(\lambda+1)} \int_0^\infty \int_0^\infty f(x)g(y) \left[\int_0^\infty e^{-(x+y)t} t^{(\lambda+1)-1} dt \right] dx\, dy \\
&= \frac{1}{\Gamma(\lambda+1)} \int_0^\infty t^{(\lambda+1)-1} \left(\int_0^\infty e^{-xt} f(x) dx \right) \\
&\quad \times \left(\int_0^\infty e^{-ty} g(y) dy \right) dt \\
&= \frac{1}{\Gamma(\lambda+1)} \int_0^\infty t^{(\lambda+1)-1} \left(\int_0^\infty e^{-xt} f(x) dx \right) \\
&\quad \times \left(\int_0^\infty t^{-1} e^{-ty} g'(y) dy \right) dt \\
&= \frac{1}{\lambda \Gamma(\lambda)} \int_0^\infty \int_0^\infty f(x) g'(y) \left[\int_0^\infty e^{-(x+y)t} t^{\lambda-1} dt \right] dx\, dy \\
&= \frac{1}{\lambda} I_1.
\end{aligned}
\tag{3.11}
$$

Then, by (3.7), we deduce (3.9).

For $a = 0$ in (3.9), we obtain (3.10). For any $0 < \varepsilon < q\lambda_2$, we set

$$
\widetilde{f}(x) := \begin{cases} 0, & 0 < x \le 1, \\ x^{\lambda_1 - \frac{\varepsilon}{p} - 1}, & x > 1; \end{cases}
\qquad
\widetilde{g}(y) := \begin{cases} 0, & 0 < y \le 1, \\ y^{\lambda_2 - \frac{\varepsilon}{q}}, & y > 1. \end{cases}
$$

We obtain that $\widetilde{g}(y) = o(1)$ $(y \to 0^+)$,

$$
\begin{aligned}
0 &< \int_0^\infty e^{-ty} \widetilde{g}(y) dy = \int_1^\infty e^{-ty} y^{\lambda_2 - \frac{\varepsilon}{q}} dy \\
&< \frac{1}{t^{\lambda_2 - \frac{\varepsilon}{q} + 1}} \int_0^\infty e^{-u} u^{(\lambda_2 - \frac{\varepsilon}{q} + 1) - 1} du \\
&= \frac{1}{t^{\lambda_2 - \frac{\varepsilon}{q} + 1}} \Gamma\left(\lambda_2 - \frac{\varepsilon}{q} + 1 \right) < \infty,
\end{aligned}
$$

$\widetilde{g}'(y) = 0 \ (0 < y < 1)$, and

$$\widetilde{g}'(y) = \left(\lambda_2 - \frac{\varepsilon}{q}\right) y^{\lambda_2 - \frac{\varepsilon}{q} - 1} \ (y > 1).$$

If there exists a positive constant $M \ (\leq \frac{1}{\lambda} B(\lambda_1, \lambda_2))$ such that (3.10) is valid when replacing

$$\frac{1}{\lambda} B(\lambda_1, \lambda_2)$$

by M, then in particular, by replacing $f(x), g(y)$, and $g'(y)$ by $\widetilde{f}(x), \widetilde{g}(y)$, and $\widetilde{g}'(y)$, respectively, we obtain that

$$\widetilde{I} := \int_0^\infty \int_0^\infty \frac{\widetilde{f}(x)\widetilde{g}(y)}{(x+y)^{\lambda+1}} dx\, dy$$

$$< M \left[\int_0^\infty x^{p(1-\lambda_1)-1} \widetilde{f}^p(x) dx\right]^{\frac{1}{p}} \left[\int_0^\infty y^{q(1-\lambda_2)-1} \widetilde{g}'^q(y) dy\right]^{\frac{1}{q}}$$

$$= M \left(\lambda_2 - \frac{\varepsilon}{q}\right) \left(\int_1^\infty x^{-\varepsilon-1} dx\right)^{\frac{1}{p}} \left(\int_1^\infty y^{-\varepsilon-1} dy\right)^{\frac{1}{q}}$$

$$= \frac{M}{\varepsilon} \left(\lambda_2 - \frac{\varepsilon}{q}\right).$$

In view of Fubini's theorem (cf. Ref. [119]), setting $u = \frac{y}{x}$, it follows that

$$\widetilde{I} := \int_1^\infty x^{\lambda_1 - \frac{\varepsilon}{p} - 1} \left[\int_1^\infty \frac{y^{\lambda_2 - \frac{\varepsilon}{q}}}{(x+y)^{\lambda+1}} dy\right] dx$$

$$= \int_1^\infty x^{-\varepsilon-1} \left[\int_{\frac{1}{x}}^\infty \frac{u^{\lambda_2 - \frac{\varepsilon}{q}}}{(1+u)^{\lambda+1}} du\right] dx$$

$$= \int_1^\infty x^{-\varepsilon-1} \left[\int_{\frac{1}{x}}^1 \frac{u^{\lambda_2 - \frac{\varepsilon}{q}}}{(1+u)^{\lambda+1}} du\right] dx$$

$$+ \int_1^\infty x^{-\varepsilon-1} \left[\int_1^\infty \frac{u^{\lambda_2 - \frac{\varepsilon}{q}}}{(1+u)^{\lambda+1}} du \right] dx$$

$$= \int_0^1 \left(\int_{\frac{1}{u}}^\infty x^{-\varepsilon-1} dx \right) \frac{u^{\lambda_2 - \frac{\varepsilon}{q}} du}{(1+u)^{\lambda+1}} + \frac{1}{\varepsilon} \int_1^\infty \frac{u^{\lambda_2 - \frac{\varepsilon}{q}} du}{(1+u)^{\lambda+1}}$$

$$= \frac{1}{\varepsilon} \left[\int_0^1 \frac{u^{\lambda_2 + \frac{\varepsilon}{p}}}{(1+u)^{\lambda+1}} du + \int_1^\infty \frac{u^{\lambda_2 - \frac{\varepsilon}{q}}}{(1+u)^{\lambda+1}} du \right].$$

In virtue of the above results, we obtain

$$\int_0^1 \frac{u^{\lambda_2 + \frac{\varepsilon}{p}}}{(1+u)^{\lambda+1}} du + \int_1^\infty \frac{u^{\lambda_2 - \frac{\varepsilon}{q}}}{(1+u)^{\lambda+1}} du = \varepsilon \tilde{I} < M \left(\lambda_2 - \frac{\varepsilon}{q} \right).$$

For $\varepsilon \to 0^+$ in the above inequality, in view of the continuity of the beta function, we deduce that

$$M\lambda_2 \geq \int_0^\infty \frac{u^{(\lambda_2+1)-1} du}{(1+u)^{\lambda+1}} = B(\lambda_1, \lambda_2 + 1) = \frac{\lambda_2}{\lambda} B(\lambda_1, \lambda_2),$$

namely,

$$M \geq \frac{1}{\lambda} B(\lambda_1, \lambda_2),$$

from which it follows that

$$M = \frac{1}{\lambda} B(\lambda_1, \lambda_2)$$

is the best possible constant factor in (3.10).

This completes the proof of the theorem. □

Remark 3.5.

(i) We set

$$\widehat{\lambda}_1 := \lambda_1 + \frac{a}{p} = \frac{\lambda - \lambda_2}{p} + \frac{\lambda_1}{q} \quad \text{and}$$

$$\widehat{\lambda}_2 := \lambda_2 + \frac{a}{q} = \frac{\lambda - \lambda_1}{q} + \frac{\lambda_2}{p}.$$

It follows that $\widehat{\lambda}_1 + \widehat{\lambda}_2 = \lambda$. We obtain that

$$0 < \widehat{\lambda}_1 < \frac{\lambda}{p} + \frac{\lambda}{q} = \lambda \quad \text{and} \quad 0 < \widehat{\lambda}_2 = \lambda - \widehat{\lambda}_1 < \lambda,$$

and then, $B(\widehat{\lambda}_1, \widehat{\lambda}_2) \in \mathbf{R}_+$. So, we rewrite (3.9) as follows:

$$\begin{aligned}
I &= \int_0^\infty \int_0^\infty \frac{f(x)g(y)}{(x+y)^{\lambda+1}} dx\, dy \\
&< \frac{1}{\lambda} B^{\frac{1}{p}}(\lambda_2, \lambda - \lambda_2) B^{\frac{1}{q}}(\lambda_1, \lambda - \lambda_1) \\
&\quad \times \left[\int_0^\infty x^{p(1-\widehat{\lambda}_1)-1} f^p(x) dx \right]^{\frac{1}{p}} \\
&\quad \times \left[\int_0^\infty y^{q(1-\widehat{\lambda}_2)-1} g'^q(y) dy \right]^{\frac{1}{q}}. \quad (3.12)
\end{aligned}$$

(ii) For $0 < p < 1$ ($q < 0$), by the reverse Hölder inequality, we similarly (cf. Ref. [120]) obtain the reverses of (3.7), (3.9), (3.10), and (3.12).

Theorem 3.6. *For $p > 1$, if the constant factor*

$$\frac{1}{\lambda} B^{\frac{1}{p}}(\lambda_2, \lambda - \lambda_2) B^{\frac{1}{q}}(\lambda_1, \lambda - \lambda_1)$$

in (3.9) (or (3.12)) is the best possible, then $a = 0$, namely, $\lambda_1 + \lambda_2 = \lambda$.

Proof. By Hölder's inequality (cf. Ref. [120]), we obtain

$$\begin{aligned}
B(\widehat{\lambda}_1, \widehat{\lambda}_2) &= \int_0^\infty \frac{u^{\widehat{\lambda}_1-1}}{(1+u)^\lambda} du = \int_0^\infty \frac{u^{\frac{\lambda-\lambda_2}{p}+\frac{\lambda_1}{q}-1}}{(1+u)^\lambda} du \\
&= \int_0^\infty \frac{1}{(1+u)^\lambda} \left(u^{\frac{\lambda-\lambda_2-1}{p}} \right) \left(u^{\frac{\lambda_1-1}{q}} \right) du \\
&\leq \left[\int_0^\infty \frac{u^{\lambda-\lambda_2-1}}{(1+u)^\lambda} du \right]^{\frac{1}{p}} \left[\int_0^\infty \frac{u^{\lambda_1-1}}{(1+u)^\lambda} du \right]^{\frac{1}{q}} \\
&= B^{\frac{1}{p}}(\lambda_2, \lambda - \lambda_2) B^{\frac{1}{q}}(\lambda_1, \lambda - \lambda_1). \quad (3.13)
\end{aligned}$$

By (3.10) (for $\lambda_i = \widehat{\lambda}_i$ $(i = 1, 2)$), since

$$\frac{1}{\lambda} B^{\frac{1}{p}}(\lambda_2, \lambda - \lambda_2) B^{\frac{1}{q}}(\lambda_1, \lambda - \lambda_1)$$

is the best possible constant factor in (3.12), we have

$$\frac{1}{\lambda} B^{\frac{1}{p}}(\lambda_2, \lambda - \lambda_2) B^{\frac{1}{q}}(\lambda_1, \lambda - \lambda_1) \leq \frac{1}{\lambda} B(\widehat{\lambda}_1, \widehat{\lambda}_2) \ (\in \mathbf{R}_+),$$

namely,

$$B^{\frac{1}{p}}(\lambda_2, \lambda - \lambda_2) B^{\frac{1}{q}}(\lambda_1, \lambda - \lambda_1) \leq B(\widehat{\lambda}_1, \widehat{\lambda}_2).$$

It follows that (3.13) retains the form of equality.

We observe that (3.13) retains the form of equality if and only if there exist constants A and B such that they are not both zero and (cf. Ref. [120])

$$Au^{\lambda - \lambda_2 - 1} = Bu^{\lambda_1 - 1} \quad \text{a.e. in } \mathbf{R}_+.$$

Assuming that $A \neq 0$, we have

$$u^{\lambda - \lambda_1 - \lambda_2} = \frac{B}{A} \quad \text{a.e. in } \mathbf{R}_+.$$

It follows that $a = \lambda - \lambda_1 - \lambda_2 = 0$, namely, $\lambda_1 + \lambda_2 = \lambda$.

This completes the proof of the theorem. $\qquad\qquad\square$

Theorem 3.7. *For $p > 1$, the following statements, (i), (ii), (iii), and (iv), are equivalent:*

(i) both $B^{\frac{1}{p}}(\lambda_2, \lambda - \lambda_2) B^{\frac{1}{q}}(\lambda_1, \lambda - \lambda_1)$ and

$$B\left(\frac{\lambda - \lambda_2}{p} + \frac{\lambda_1}{q}, \frac{\lambda - \lambda_1}{q} + \frac{\lambda_1}{p}\right)$$

   are independent of p and q;

(ii) the following equality holds:

$$B^{\frac{1}{p}}(\lambda_2, \lambda - \lambda_2) B^{\frac{1}{q}}(\lambda_1, \lambda - \lambda_1)$$
$$= B\left(\frac{\lambda - \lambda_2}{p} + \frac{\lambda_1}{q}, \frac{\lambda - \lambda_1}{q} + \frac{\lambda_1}{p}\right);$$

(iii) $\lambda_1 + \lambda_2 = \lambda$;
(iv) the constant factor

$$\frac{1}{\lambda} B^{\frac{1}{p}}(\lambda_2, \lambda - \lambda_2) B^{\frac{1}{q}}(\lambda_1, \lambda - \lambda_1)$$

is the best possible in (3.9).

Proof.   (i) $\Rightarrow$ (ii): In view of the continuity of the beta function, we derive that

$$B^{\frac{1}{p}}(\lambda_2, \lambda - \lambda_2) B^{\frac{1}{q}}(\lambda_1, \lambda - \lambda_1)$$
$$= \lim_{p \to 1^+} \lim_{q \to \infty} B^{\frac{1}{p}}(\lambda_2, \lambda - \lambda_2) B^{\frac{1}{q}}(\lambda_1, \lambda - \lambda_1)$$
$$= B(\lambda_2, \lambda - \lambda_2),$$
$$\times B\left(\frac{\lambda - \lambda_2}{p} + \frac{\lambda_1}{q}, \frac{\lambda - \lambda_1}{q} + \frac{\lambda_2}{p}\right)$$
$$= \lim_{p \to 1^+} \lim_{q \to \infty} B\left(\frac{\lambda - \lambda_2}{p} + \frac{\lambda_1}{q}, \frac{\lambda - \lambda_1}{q} + \frac{\lambda_2}{p}\right)$$
$$= B(\lambda_2, \lambda - \lambda_2) = B^{\frac{1}{p}}(\lambda_2, \lambda - \lambda_2) B^{\frac{1}{q}}(\lambda_1, \lambda - \lambda_1).$$

(ii) $\Rightarrow$ (iii): In view of this expression, (3.13) retains the form of equality. By the proof of Theorem 3.6, we have $\lambda_1 + \lambda_2 = \lambda$.

(iii) $\Rightarrow$ (iv): If $\lambda_1 + \lambda_2 = \lambda$, then by Theorem 3.4, the constant factor

$$\frac{1}{\lambda} B^{\frac{1}{p}}(\lambda_2, \lambda - \lambda_2) B^{\frac{1}{q}}(\lambda_1, \lambda - \lambda_1) \ \left(= \frac{1}{\lambda} B(\lambda_1, \lambda_2)\right)$$

in (3.9) is the best possible.

(iv) $\Rightarrow$ (i): By Theorem 3.6, we have $\lambda_1 + \lambda_2 = \lambda$, and then,

$$B^{\frac{1}{p}}(\lambda_2, \lambda - \lambda_2) B^{\frac{1}{q}}(\lambda_1, \lambda - \lambda_1) = B(\lambda_1, \lambda_2),$$
$$B\left(\frac{\lambda - \lambda_2}{p} + \frac{\lambda_1}{q}, \frac{\lambda - \lambda_1}{q} + \frac{\lambda_2}{p}\right) = B(\lambda_1, \lambda_2).$$

Both of them are independent of p and q.

Hence, the statements (i), (ii), (iii), and (iv) are equivalent. This completes the proof of the theorem. $\qquad\square$

3.3   Equivalent Form and Some Particular Inequalities

Theorem 3.8. *For $p > 1$, inequality (3.9) is equivalent to the following Hilbert-type integral inequality involving one derivative function:*

$$J := \left\{ \int_0^\infty x^{q(\lambda_1 + a) - a - 1} \left[\int_0^\infty \frac{g(y)}{(x+y)^{\lambda+1}} dy \right]^q dx \right\}^{\frac{1}{q}}$$

$$< \frac{1}{\lambda} B^{\frac{1}{p}}(\lambda_2, \lambda - \lambda_2) B^{\frac{1}{q}}(\lambda_1, \lambda - \lambda_1)$$

$$\times \left[\int_0^\infty y^{q(1-\lambda_2) - a - 1} g'^q(y) dy \right]^{\frac{1}{q}}. \tag{3.14}$$

In particular, for $\lambda_1 + \lambda_2 = \lambda$, we reduce (3.14) to the equivalent form of (3.10) as follows:

$$\left\{ \int_0^\infty x^{q\lambda_1 - 1} \left[\int_0^\infty \frac{g(y)}{(x+y)^{\lambda+1}} dy \right]^q dx \right\}^{\frac{1}{q}}$$

$$< \frac{1}{\lambda} B(\lambda_1, \lambda_2) \left[\int_0^\infty y^{q(1-\lambda_2) - 1} g'^q(y) dy \right]^{\frac{1}{q}}, \tag{3.15}$$

where the constant factor $\frac{1}{\lambda} B(\lambda_1, \lambda_2)$ is the best possible.

Proof.   Suppose that (3.14) is valid. By Hölder's inequality, we have

$$I = \int_0^\infty \left[x^{\frac{1}{q} - \lambda_1 - \frac{a}{p}} f(x) \right] \left[x^{\frac{-1}{q} + \lambda_1 + \frac{a}{p}} \int_0^\infty \frac{g(y)}{(x+y)^{\lambda+1}} dy \right] dx$$

$$\leq \left[\int_0^\infty x^{p(1-\lambda_1) - a - 1} f^p(x) dx \right]^{\frac{1}{p}} J. \tag{3.16}$$

Then, by (3.14), we have (3.9).

On the other hand, assuming that (3.9) is valid, we set

$$f(x) := x^{q(\lambda_1+a)-a-1} \left[\int_0^\infty \frac{g(y)}{(x+y)^{\lambda+1}} dy \right]^{q-1}, \quad x \in \mathbf{R}_+.$$

If $J = 0$, then (3.14) is naturally valid; if $J = \infty$, then it is impossible to make (3.14) valid, namely, $J < \infty$. Suppose that $0 < J < \infty$. By (3.9), we have

$$0 < \int_0^\infty x^{p(1-\lambda_1)-a-1} f^p(x) dx = J^q = I$$

$$< \frac{1}{\lambda} B^{\frac{1}{p}}(\lambda_2, \lambda - \lambda_2) B^{\frac{1}{q}}(\lambda_1, \lambda - \lambda_1) J^{q-1}$$

$$\times \left[\int_0^\infty y^{q(1-\lambda_2)-a-1} g'^q(y) dy \right]^{\frac{1}{q}} < \infty,$$

$$J = \left[\int_0^\infty x^{p(1-\lambda_1)-a-1} f^p(x) dx \right]^{\frac{1}{q}}$$

$$< \frac{1}{\lambda} B^{\frac{1}{p}}(\lambda_2, \lambda - \lambda_2) B^{\frac{1}{q}}(\lambda_1, \lambda - \lambda_1)$$

$$\times \left[\int_0^\infty y^{q(1-\lambda_2)-a-1} g'^q(y) dy \right]^{\frac{1}{q}},$$

namely, (3.14) follows, which is equivalent to (3.9).

The constant factor $\frac{1}{\lambda} B(\lambda_1, \lambda_2)$ is the best possible in (3.15). Otherwise, by (3.16) (for $a = 0$), we would reach a contradiction that the constant factor in (3.10) is not the best possible.

This completes the proof of the theorem. $\quad\square$

Replacing x by $\frac{1}{x}$, then replacing $x^{\lambda-1} f(\frac{1}{x})$ by $f(x)$ in (3.9) and (3.14), we deduce the following corollary.

Corollary 3.9. *The following Hilbert-type integral inequalities with a nonhomogeneous kernel involving one derivative function are equivalent:*

$$\int_0^\infty \int_0^\infty \frac{f(x)g(y)}{(1+xy)^{\lambda+1}} dx\, dy < \frac{1}{\lambda} B^{\frac{1}{p}}(\lambda_2, \lambda - \lambda_2) B^{\frac{1}{q}}(\lambda_1, \lambda - \lambda_1)$$

$$\times \left[\int_0^\infty x^{p(\lambda_1-\lambda)+a-1} f^p(x) dx \right]^{\frac{1}{p}}$$

$$\times \left[\int_0^\infty y^{q(1-\lambda_2)-a-1} g'^q(y) dy \right]^{\frac{1}{q}}, \tag{3.17}$$

$$\left\{ \int_0^\infty x^{q(\lambda-\lambda_1-a+1)+a-1} \left[\int_0^\infty \frac{g(y)}{(1+xy)^{\lambda+1}} dy \right]^q dx \right\}^{\frac{1}{q}}$$

$$< \frac{1}{\lambda} B^{\frac{1}{p}}(\lambda_2, \lambda - \lambda_2) B^{\frac{1}{q}}(\lambda_1, \lambda - \lambda_1)$$

$$\times \left[\int_0^\infty y^{q(1-\lambda_2)-a-1} g'^q(y) dy \right]^{\frac{1}{q}}. \tag{3.18}$$

Moreover, $\lambda_1 + \lambda_2 = \lambda$ if and only if the constant factor

$$\frac{1}{\lambda} B^{\frac{1}{p}}(\lambda_2, \lambda - \lambda_2) B^{\frac{1}{q}}(\lambda_1, \lambda - \lambda_1)$$

in (3.17) and (3.18) is the best possible.

For $\lambda_1 + \lambda_2 = \lambda$, we have the following equivalent inequalities with the best possible constant factor $\frac{1}{\lambda} B(\lambda_1, \lambda_2)$:

$$\int_0^\infty \int_0^\infty \frac{f(x)g(y)}{(1+xy)^{\lambda+1}} dx\, dy$$

$$< \frac{1}{\lambda} B(\lambda_1, \lambda_2) \left(\int_0^\infty x^{-p\lambda_2-1} f^p(x) dx \right)^{\frac{1}{p}}$$

$$\times \left[\int_0^\infty y^{q(1-\lambda_2)-1} g'^q(y) dy \right]^{\frac{1}{q}}, \tag{3.19}$$

$$\left\{ \int_0^\infty x^{q(1+\lambda_2)-1} \left[\int_0^\infty \frac{g(y)}{(1+xy)^{\lambda+1}} dy \right]^q dx \right\}^{\frac{1}{q}}$$

$$< \frac{1}{\lambda} B(\lambda_1, \lambda_2) \left[\int_0^\infty y^{q(1-\lambda_2)-1} g'^q(y) dy \right]^{\frac{1}{q}}. \tag{3.20}$$

Remark 3.10. For

$$\lambda_1 = \frac{\lambda}{r} \quad \text{and} \quad \lambda_2 = \frac{\lambda}{s} \quad \left(r > 1, \frac{1}{r} + \frac{1}{s} = 1 \right)$$

in (3.10), (3.15), (3.19), and (3.20), we have the following two couples of equivalent integral inequalities:

$$\int_0^\infty \int_0^\infty \frac{f(x)g(y)}{(x+y)^{\lambda+1}} dx\, dy < \frac{1}{\lambda} B\left(\frac{\lambda}{r}, \frac{\lambda}{s}\right)$$

$$\times \left[\int_0^\infty x^{p(1-\frac{\lambda}{r})-1} f^p(x) dx\right]^{\frac{1}{p}}$$

$$\times \left[\int_0^\infty y^{q(1-\frac{\lambda}{s})-1} g'^q(y) dy\right]^{\frac{1}{q}}, \qquad (3.21)$$

$$\left\{\int_0^\infty x^{\frac{q\lambda}{r}-1} \left[\int_0^\infty \frac{g(y)}{(x+y)^{\lambda+1}} dy\right]^q dx\right\}^{\frac{1}{q}}$$

$$< \frac{1}{\lambda} B\left(\frac{\lambda}{r}, \frac{\lambda}{s}\right) \left[\int_0^\infty y^{q(1-\frac{\lambda}{s})-1} g'^q(y) dy\right]^{\frac{1}{q}}; \qquad (3.22)$$

$$\int_0^\infty \int_0^\infty \frac{f(x)g(y)}{(1+xy)^{\lambda+1}} dx\, dy < \frac{1}{\lambda} B\left(\frac{\lambda}{r}, \frac{\lambda}{s}\right)$$

$$\times \left(\int_0^\infty x^{-\frac{p\lambda}{s}-1} f^p(x) dx\right)^{\frac{1}{p}}$$

$$\times \left[\int_0^\infty y^{q(1-\frac{\lambda}{s})-1} g'^q(y) dy\right]^{\frac{1}{q}}, \qquad (3.23)$$

$$\left\{\int_0^\infty x^{q(1+\frac{\lambda}{s})-1} \left[\int_0^\infty \frac{g(y)}{(1+xy)^{\lambda+1}} dy\right]^q dx\right\}^{\frac{1}{q}}$$

$$< \frac{1}{\lambda} B\left(\frac{\lambda}{r}, \frac{\lambda}{s}\right) \left[\int_0^\infty y^{q(1-\frac{\lambda}{s})-1} g'^q(y) dy\right]^{\frac{1}{q}}, \qquad (3.24)$$

where the constant factor

$$\frac{1}{\lambda} B\left(\frac{\lambda}{r}, \frac{\lambda}{s}\right)$$

is the best possible.

In particular:

(i) For $\lambda = 1, r = q$, and $s = p$, we have

$$\int_0^\infty \int_0^\infty \frac{f(x)g(y)}{(x+y)^2} dx\, dy < \frac{\pi}{\sin(\pi/p)} \left(\int_0^\infty f^p(x) dx \right)^{\frac{1}{p}}$$

$$\times \left(\int_0^\infty g'^q(y) dy \right)^{\frac{1}{q}}, \tag{3.25}$$

$$\left\{ \int_0^\infty \left[\int_0^\infty \frac{g(y)}{(x+y)^2} dy \right]^q dx \right\}^{\frac{1}{q}} < \frac{\pi}{\sin(\pi/p)}$$

$$\times \left(\int_0^\infty g'^q(y) dy \right)^{\frac{1}{q}}; \tag{3.26}$$

$$\int_0^\infty \int_0^\infty \frac{f(x)g(y)}{(1+xy)^2} dx\, dy < \frac{\pi}{\sin(\pi/p)} \left(\int_0^\infty x^{-2} f^p(x) dx \right)^{\frac{1}{p}}$$

$$\times \left(\int_0^\infty g'^q(y) dy \right)^{\frac{1}{q}}, \tag{3.27}$$

$$\left\{ \int_0^\infty x^{2(q-1)} \left[\int_0^\infty \frac{g(y) dy}{(1+xy)^2} \right]^q dx \right\}^{\frac{1}{q}}$$

$$< \frac{\pi}{\sin(\pi/p)} \left(\int_0^\infty g'^q(y) dy \right)^{\frac{1}{q}}. \tag{3.28}$$

(ii) For $\lambda = 1, r = p$, and $s = q$, we have the following dual forms of (3.25)–(3.28):

$$\int_0^\infty \int_0^\infty \frac{f(x)g(y)}{(x+y)^2} dx\, dy < \frac{\pi}{\sin(\pi/p)}$$

$$\times \left(\int_0^\infty x^{p-2} f^p(x) dx \right)^{\frac{1}{p}} \left(\int_0^\infty y^{q-2} g'^q(y) dy \right)^{\frac{1}{q}}, \tag{3.29}$$

$$\left\{ \int_0^\infty x^{q-2} \left[\int_0^\infty \frac{g(y) dy}{(x+y)^2} \right]^q dx \right\}^{\frac{1}{q}}$$

$$< \frac{\pi}{\sin(\pi/p)} \left(\int_0^\infty y^{q-2} g'^q(y) dy \right)^{\frac{1}{q}}; \tag{3.30}$$

$$\int_0^\infty \int_0^\infty \frac{f(x)g(y)}{(1+xy)^2} dx\, dy < \frac{\pi}{\sin(\pi/p)}$$

$$\times \left(\int_0^\infty x^{-p} f^p(x) dx \right)^{\frac{1}{p}} \left(\int_0^\infty y^{q-2} g'^q(y) dy \right)^{\frac{1}{q}}, \quad (3.31)$$

$$\left\{ \int_0^\infty x^q \left[\int_0^\infty \frac{g(y) dy}{(1+xy)^2} \right]^q dx \right\}^{\frac{1}{q}}$$

$$< \frac{\pi}{\sin(\pi/p)} \left(\int_0^\infty y^{q-2} g'^q(y) dy \right)^{\frac{1}{q}}. \quad (3.32)$$

(iii) For $p = q = 2$, both (3.25) and (3.29) reduce to

$$\int_0^\infty \int_0^\infty \frac{f(x)g(y)}{(x+y)^2} dx\, dy < \pi \left(\int_0^\infty f^2(x) dx \int_0^\infty g'^2(y) dy \right)^{\frac{1}{2}} \quad (3.33)$$

and both (3.26) and (3.30) reduce to the following equivalent inequality of (3.33):

$$\left\{ \int_0^\infty \left[\int_0^\infty \frac{g(y)}{(x+y)^2} dy \right]^2 dx \right\}^{\frac{1}{2}} < \pi \left(\int_0^\infty g'^2(y) dy \right)^{\frac{1}{2}}; \quad (3.34)$$

both (3.27) and (3.31) reduce to

$$\int_0^\infty \int_0^\infty \frac{f(x)g(y)}{(1+xy)^2} dx\, dy < \pi \left(\int_0^\infty x^{-2} f^2(x) dx \int_0^\infty g'^2(y) dy \right)^{\frac{1}{2}}$$

$$(3.35)$$

and both (3.28) and (3.32) reduce to the following equivalent inequality of (3.35):

$$\left\{ \int_0^\infty x^2 \left[\int_0^\infty \frac{g(y)}{(1+xy)^2} dy \right]^2 dx \right\}^{\frac{1}{2}} < \pi \left(\int_0^\infty g'^2(y) dy \right)^{\frac{1}{2}}. \quad (3.36)$$

3.4   Operator Expressions

For $p > 1$, we set

$$\varphi(x) := x^{p(1-\lambda_1)-a-1} \quad \text{and} \quad \psi(y) := y^{q(1-\lambda_2)-a-1},$$

wherefrom

$$\varphi^{1-q}(x) = x^{q(\lambda_1+a)-a-1} (x, y \in \mathbf{R}_+).$$

Define the real normed linear spaces $L_{p,\varphi}(\mathbf{R}_+), L_{q,\psi}(\mathbf{R}_+)$, and $L_{q,\varphi^{1-q}}(\mathbf{R}_+)$ as in Chapter 2.

Assuming that $g(y)$ is a nonnegative increasing function in $\mathbf{R}_+$ such that $g'(y) \geq 0$ and

$$g \in \widetilde{L}(\mathbf{R}_+) := \{g = g(y); g(y) = o(1)(y \to 0^+), e^{-ty}g(y) \in L(\mathbf{R}_+)\}$$

and setting

$$h(x) := \int_0^\infty \frac{g(y)}{(x+y)^{\lambda+1}} dy, \quad x \in \mathbf{R}_+,$$

we can rewrite (3.14) as follows:

$$\|h\|_{q,\varphi^{1-q}} \leq \frac{1}{\lambda} B^{\frac{1}{p}}(\lambda_2, \lambda - \lambda_2) B^{\frac{1}{q}}(\lambda_1, \lambda - \lambda_1) \|g'\|_{q,\psi} < \infty,$$

namely, $h \in L_{q,\varphi^{1-q}}(\mathbf{R}_+)$.

Definition 3.11. For $p > 1$, define a Hilbert-type operator

$$T : \widetilde{L}(\mathbf{R}_+) \to L_{q,\varphi^{1-q}}(\mathbf{R}_+)$$

as follows.
For any $g \in \widetilde{L}(\mathbf{R}_+)$, there exists a unique representation $h \in L_{q,\varphi^{1-q}}(\mathbf{R}_+)$ satisfying

$$(Tg)(x) = h(x)$$

for any $x \in \mathbf{R}_+$. Define the formal inner product of $f \in L_{p,\varphi}(\mathbf{R}_+)$ and Tg and the norm of T as follows:

$$(f, Tg) := \int_0^\infty f(x) \left[\int_0^\infty \frac{g(y)}{(x+y)^{\lambda+1}} dy \right] dx = I,$$

$$\|T\| = \sup_{g(\neq\theta)\in\widetilde{L}(\mathbf{R}_+)} \frac{\|Tg\|_{q,\varphi^{1-q}}}{\|g'\|_{q,\psi}}.$$

By Theorem 3.8, we obtain the following theorem.

Theorem 3.12. *For* $p > 1$, *if* $f \in L_{p,\varphi}(\mathbf{R}_+), g \in \tilde{L}(\mathbf{R}_+)$, $\|f\|_{p,\varphi}, \|g'\|_{q,\psi} > 0$, *then we have the following equivalent inequalities:*

$$(f, Tg) < \frac{1}{\lambda}B^{\frac{1}{p}}(\lambda_2, \lambda - \lambda_2)B^{\frac{1}{q}}(\lambda_1, \lambda - \lambda_1)\|f\|_{p,\varphi}\|g'\|_{q,\psi}, \quad (3.37)$$

$$\|Tg\|_{q,\varphi^{1-q}} < \frac{1}{\lambda}B^{\frac{1}{p}}(\lambda_2, \lambda - \lambda_2)B^{\frac{1}{q}}(\lambda_1, \lambda - \lambda_1)\|g'\|_{q,\psi}. \quad (3.38)$$

Moreover, $\lambda_1 + \lambda_2 = \lambda$ if and only if the constant factor

$$\frac{1}{\lambda}B^{\frac{1}{p}}(\lambda_2, \lambda - \lambda_2)B^{\frac{1}{q}}(\lambda_1, \lambda - \lambda_1)$$

in (3.37) and (3.38) is the best possible, namely,

$$\|T\| = \frac{1}{\lambda}B(\lambda_1, \lambda_2).$$

For $p > 1$, we also set

$$\phi(x) := x^{p(\lambda_1 - \lambda) + a - 1},$$

wherefrom

$$\phi^{1-q}(x) = x^{q(\lambda - \lambda_1 - a + 1) + a - 1} \ (x \in \mathbf{R}_+).$$

Define the real normed linear spaces $L_{p,\phi}(\mathbf{R}_+)$ and $L_{q,\phi^{1-q}}(\mathbf{R}_+)$ as in Chapter 2.

Assuming that $g(y)$ is a nonnegative increasing function in $\mathbf{R}_+$, $g \in \tilde{L}(\mathbf{R}_+)$, and setting

$$H(x) := \int_0^\infty \frac{g(y)}{(1 + xy)^{\lambda + 1}}dy, \ x \in \mathbf{R}_+,$$

we can rewrite (3.18) as follows:

$$\|H\|_{q,\phi^{1-q}} \leq \frac{1}{\lambda}B^{\frac{1}{p}}(\lambda_2, \lambda - \lambda_2)B^{\frac{1}{q}}(\lambda_1, \lambda - \lambda_1)\|g'\|_{q,\psi} < \infty,$$

namely, $H \in L_{q,\phi^{1-q}}(\mathbf{R}_+)$.

Definition 3.13. For $p > 1$, define a Hilbert-type operator

$$T_1 : \tilde{L}(\mathbf{R}_+) \to L_{q,\phi^{1-q}}(\mathbf{R}_+)$$

as follows.

For any $g \in \tilde{L}(\mathbf{R}_+)$, there exists a unique representation $H \in L_{q,\phi^{1-q}}(\mathbf{R}_+)$ satisfying

$$(T_1 g)(x) = H(x)$$

for any $x \in \mathbf{R}_+$. Define the formal inner product of $f \in L_{p,\phi}(\mathbf{R}_+)$ and $T_1 g$ and the norm of T_1 as follows:

$$(f, T_1 g) := \int_0^\infty f(x) \left[\int_0^\infty \frac{g(y)}{(1+xy)^{\lambda+1}} dy \right] dx = I,$$

$$\|T_1\| = \sup_{g(\neq\theta)\in\tilde{L}(\mathbf{R}_+)} \frac{\|Tg\|_{q,\phi^{1-q}}}{\|g'\|_{q,\psi}}.$$

By Corollary 3.9, we have the following theorem.

Theorem 3.14. *For $p > 1$, if $f \in L_{p,\phi}(\mathbf{R}_+), g \in \tilde{L}(\mathbf{R}_+)$, $\|f\|_{p,\phi}, \|g'\|_{q,\psi} > 0$, then we have the following equivalent inequalities:*

$$(f, T_1 g) < \frac{1}{\lambda} B^{\frac{1}{p}}(\lambda_2, \lambda - \lambda_2) B^{\frac{1}{q}}(\lambda_1, \lambda - \lambda_1)\|f\|_{p,\phi}\|g'\|_{q,\psi}, \quad (3.39)$$

$$\|Tg\|_{q,\phi^{1-q}} < \frac{1}{\lambda} B^{\frac{1}{p}}(\lambda_2, \lambda - \lambda_2) B^{\frac{1}{q}}(\lambda_1, \lambda - \lambda_1)\|g'\|_{q,\psi}. \quad (3.40)$$

Moreover, $\lambda_1 + \lambda_2 = \lambda$ if and only if the constant factor

$$\frac{1}{\lambda} B^{\frac{1}{p}}(\lambda_2, \lambda - \lambda_2) B^{\frac{1}{q}}(\lambda_1, \lambda - \lambda_1)$$

in (3.39) and (3.40) is the best possible, namely,

$$\|T_1\| = \frac{1}{\lambda} B(\lambda_1, \lambda_2).$$

3.5 The Case of Reverses

Theorem 3.15. *For* $0 < p < 1$ $(q < 0)$*, we have the following reverse Hilbert-type integral inequality involving one derivative function:*

$$I := \int_0^\infty \int_0^\infty \frac{f(x)g(y)}{(x+y)^{\lambda+1}} dx\, dy$$

$$> \frac{1}{\lambda} B^{\frac{1}{p}}(\lambda_2, \lambda - \lambda_2) B^{\frac{1}{q}}(\lambda_1, \lambda - \lambda_1)$$

$$\times \left[\int_0^\infty x^{p(1-\lambda_1)-a-1} f^p(x) dx \right]^{\frac{1}{p}}$$

$$\times \left[\int_0^\infty y^{q(1-\lambda_2)-a-1} g'^q(y) dy \right]^{\frac{1}{q}}. \tag{3.41}$$

In particular, for $\lambda_1 + \lambda_2 = \lambda$ (or $a = 0$), we reduce (3.41) to the following:

$$\int_0^\infty \int_0^\infty \frac{f(x)g(y)}{(x+y)^{\lambda+1}} dx\, dy > \frac{1}{\lambda} B(\lambda_1, \lambda_2)$$

$$\times \left[\int_0^\infty x^{p(1-\lambda_1)-1} f^p(x) dx \right]^{\frac{1}{p}} \left[\int_0^\infty y^{q(1-\lambda_2)-1} g'^q(y) dy \right]^{\frac{1}{q}}, \tag{3.42}$$

where the constant factor $\frac{1}{\lambda} B(\lambda_1, \lambda_2)$ is the best possible.

Proof. Using (3.11) and the reverse of (3.7), we derive (3.41).

For $a = 0$ in (3.41), we obtain (3.42). For any $0 < \varepsilon < |q|\lambda_1$, we set

$$\widetilde{f}(x) := \begin{cases} 0, & 0 < x \leq 1, \\ x^{\lambda_1 - \frac{\varepsilon}{p} - 1}, & x > 1; \end{cases} \qquad \widetilde{g}(y) := \begin{cases} 0, & 0 < y \leq 1, \\ y^{\lambda_2 - \frac{\varepsilon}{q}}, & y > 1. \end{cases}$$

We obtain that $\widetilde{g}'(y) = 0$ $(0 < y < 1)$ and

$$\widetilde{g}'(y) = \left(\lambda_2 - \frac{\varepsilon}{q} \right) y^{\lambda_2 - \frac{\varepsilon}{q} - 1} \quad (y > 1).$$

If there exists a positive constant M $(\geq \frac{1}{\lambda} B(\lambda_1, \lambda_2))$ such that (3.42) is valid when replacing $\frac{1}{\lambda} B(\lambda_1, \lambda_2)$ by M, then in particular, replacing $f(x), g(y)$, and $g'(y)$ by $\widetilde{f}(x), \widetilde{g}(y)$, and $\widetilde{g}'(y)$, respectively, we have

$$
\begin{aligned}
\widetilde{I} &:= \int_0^\infty \int_0^\infty \frac{\widetilde{f}(x)\widetilde{g}(y)}{(x+y)^{\lambda+1}} dx\, dy \\
&> M \left[\int_0^\infty x^{p(1-\lambda_1)-1}\widetilde{f}^p(x)dx \right]^{\frac{1}{p}} \left[\int_0^\infty y^{q(1-\lambda_2)-1}\widetilde{g}'^q(y)dy \right]^{\frac{1}{q}} \\
&= M\left(\lambda_2 - \frac{\varepsilon}{q}\right) \int_1^\infty x^{-\varepsilon-1}dx = \frac{M}{\varepsilon}\left(\lambda_2 - \frac{\varepsilon}{q}\right).
\end{aligned}
$$

For

$$
\lambda + 1 = \left(\lambda_2 + 1 - \frac{\varepsilon}{q}\right) + \left(\lambda_1 + \frac{\varepsilon}{q}\right),
$$

replacing λ (resp. λ_2) by $\lambda+1$ (resp. $\lambda_2 + 1 - \frac{\varepsilon}{q}$) in (3.3), we have

$$
\begin{aligned}
\varpi_{\lambda+1}\left(\lambda_2 + 1 - \frac{\varepsilon}{q}, x\right) &= x^{\lambda_1+\frac{\varepsilon}{q}} \int_0^\infty \frac{y^{\lambda_2-\frac{\varepsilon}{q}}}{(x+y)^{\lambda+1}}dy \\
&= B\left(\lambda_2 + 1 - \frac{\varepsilon}{q}, \lambda_1 + \frac{\varepsilon}{q}\right) \quad (x \in \mathbf{R}_+),
\end{aligned}
$$

and then,

$$
\begin{aligned}
\widetilde{I} &\le \int_1^\infty x^{\lambda_1-\frac{\varepsilon}{p}-1} \left[\int_0^\infty \frac{y^{\lambda_2-\frac{\varepsilon}{q}}}{(x+y)^{\lambda+1}}dy \right] dx \\
&= \int_1^\infty x^{-\varepsilon-1}\varpi_{\lambda+1}\left(\lambda_2 + 1 - \frac{\varepsilon}{q}, x\right) dx \\
&= \frac{1}{\varepsilon}B(\lambda_2 + 1 - \frac{\varepsilon}{q}, \lambda_1 + \frac{\varepsilon}{q}).
\end{aligned}
$$

In virtue of the above results, we obtain

$$
B\left(\lambda_2 + 1 - \frac{\varepsilon}{q}, \lambda_1 + \frac{\varepsilon}{q}\right) \ge \varepsilon\widetilde{I} > M\left(\lambda_2 - \frac{\varepsilon}{q}\right).
$$

For $\varepsilon \to 0^+$ in the above inequality, in view of the continuity of the beta function, we deduce that

$$
M\lambda_2 \le B(\lambda_2 + 1, \lambda_1) = \frac{\lambda_2}{\lambda}B(\lambda_1, \lambda_2),
$$

namely,

$$M \leq \frac{1}{\lambda} B(\lambda_1, \lambda_2).$$

Hence,

$$M = \frac{1}{\lambda} B(\lambda_1, \lambda_2)$$

is the best possible constant factor in (3.42).

This completes the proof of the theorem. □

Remark 3.16. For $0 < p < 1$, we set

$$\widehat{\lambda}_1 := \lambda_1 + \frac{a}{p} = \frac{\lambda - \lambda_2}{p} + \frac{\lambda_1}{q} \quad \text{and}$$

$$\widehat{\lambda}_2 := \lambda_2 + \frac{a}{q} = \frac{\lambda - \lambda_1}{q} + \frac{\lambda_2}{p}.$$

It follows that $\widehat{\lambda}_1 + \widehat{\lambda}_2 = \lambda$. For

$$a = \lambda - \lambda_1 - \lambda_2 \in (-p\lambda_1, p(\lambda - \lambda_1)),$$

we derive that $0 < \widehat{\lambda}_1 < \lambda, 0 < \widehat{\lambda}_2 = \lambda - \widehat{\lambda}_1 < \lambda$, and then, $B(\widehat{\lambda}_1, \widehat{\lambda}_2) \in \mathbf{R}_+$. So, we rewrite (3.41) as follows:

$$\begin{aligned}
I = \int_0^\infty \int_0^\infty \frac{f(x)g(y)}{(x+y)^{\lambda+1}} dx\, dy \\
> \frac{1}{\lambda} B^{\frac{1}{p}}(\lambda_2, \lambda - \lambda_2) B^{\frac{1}{q}}(\lambda_1, \lambda - \lambda_1) \\
\times \left[\int_0^\infty x^{p(1-\widehat{\lambda}_1)-1} f^p(x) dx \right]^{\frac{1}{p}} \left[\int_0^\infty y^{q(1-\widehat{\lambda}_2)-1} g'^q(y) dy \right]^{\frac{1}{q}}.
\end{aligned}$$

$$(3.43)$$

Theorem 3.17. *For $0 < p < 1$, if*

$$a = \lambda - \lambda_1 - \lambda_2 \in (-p\lambda_1, p(\lambda - \lambda_1))$$

and the constant factor

$$\frac{1}{\lambda}B^{\frac{1}{p}}(\lambda_2, \lambda - \lambda_2)B^{\frac{1}{q}}(\lambda_1, \lambda - \lambda_1)$$

in (3.41) (*or* (3.43)) *is the best possible, then* $a = 0$, *namely,* $\lambda_1 + \lambda_2 = \lambda$.

Proof. By (3.42) (for $\lambda_i = \widehat{\lambda}_i$ ($i = 1, 2$)), since

$$\frac{1}{\lambda}B^{\frac{1}{p}}(\lambda_2, \lambda - \lambda_2)B^{\frac{1}{q}}(\lambda_1, \lambda - \lambda_1)$$

is the best possible constant factor in (3.43), we have

$$\frac{1}{\lambda}B^{\frac{1}{p}}(\lambda_2, \lambda - \lambda_2)B^{\frac{1}{q}}(\lambda_1, \lambda - \lambda_1) \geq \frac{1}{\lambda}B(\widehat{\lambda}_1, \widehat{\lambda}_2)(\in \mathbf{R}_+),$$

namely,

$$B^{\frac{1}{p}}(\lambda_2, \lambda - \lambda_2)B^{\frac{1}{q}}(\lambda_1, \lambda - \lambda_1) \geq B(\widehat{\lambda}_1, \widehat{\lambda}_2).$$

By the reverse Hölder inequality (cf. Ref. [120]), we obtain

$$B(\widehat{\lambda}_1, \widehat{\lambda}_2) = \int_0^\infty \frac{1}{(1+u)^\lambda}\left(u^{\frac{\lambda - \lambda_2 - 1}{p}}\right)\left(u^{\frac{\lambda_1 - 1}{q}}\right)du$$

$$\geq \left[\int_0^\infty \frac{u^{\lambda - \lambda_2 - 1}}{(1+u)^\lambda}du\right]^{\frac{1}{p}}\left[\int_0^\infty \frac{u^{\lambda_1 - 1}}{(1+u)^\lambda}du\right]^{\frac{1}{q}}$$

$$= B^{\frac{1}{p}}(\lambda_2, \lambda - \lambda_2)B^{\frac{1}{q}}(\lambda_1, \lambda - \lambda_1). \qquad (3.44)$$

It follows that (3.44) retains the form of equality.

We observe that (3.44) retains the form of equality if and only if there exist constants A and B such that they are not both zero and (cf. Ref. [120])

$$Au^{\lambda - \lambda_2 - 1} = Bu^{\lambda_1 - 1} \quad \text{a.e. in } \mathbf{R}_+.$$

Assuming that $A \neq 0$, we have

$$u^{\lambda - \lambda_1 - \lambda_2} = \frac{B}{A} \quad \text{a.e. in } \mathbf{R}_+.$$

It follows that $a = \lambda - \lambda_1 - \lambda_2 = 0$, namely, $\lambda_1 + \lambda_2 = \lambda$. This completes the proof of the theorem. $\qquad \square$

Theorem 3.18. *For $0 < p < 1$, the following statements, (i), (ii), (iii), and (iv), are equivalent:*

(i) both $B^{\frac{1}{p}}(\lambda_2, \lambda - \lambda_2)B^{\frac{1}{q}}(\lambda_1, \lambda - \lambda_1)$ and

$$B\left(\frac{\lambda - \lambda_2}{p} + \frac{\lambda_1}{q}, \frac{\lambda - \lambda_1}{q} + \frac{\lambda_1}{p}\right)$$

   are independent of p and q;

(ii) the following equality holds:

$$B^{\frac{1}{p}}(\lambda_2, \lambda - \lambda_2)B^{\frac{1}{q}}(\lambda_1, \lambda - \lambda_1)$$
$$= B\left(\frac{\lambda - \lambda_2}{p} + \frac{\lambda_1}{q}, \frac{\lambda - \lambda_1}{q} + \frac{\lambda_1}{p}\right);$$

(iii) if $a = \lambda - \lambda_1 - \lambda_2 \in (-p\lambda_1, p(\lambda - \lambda_1))$, then we have $\lambda_1 + \lambda_2 = \lambda$;

(iv) the constant factor

$$\frac{1}{\lambda}B^{\frac{1}{p}}(\lambda_2, \lambda - \lambda_2)B^{\frac{1}{q}}(\lambda_1, \lambda - \lambda_1)$$

   is the best possible in (3.41).

Proof. (i) $\Rightarrow$ (ii): In view of the continuity of the beta function, we obtain that

$$B^{\frac{1}{p}}(\lambda_2, \lambda - \lambda_2)B^{\frac{1}{q}}(\lambda_1, \lambda - \lambda_1)$$
$$= \lim_{p\to 1^-}\lim_{q\to-\infty} B^{\frac{1}{p}}(\lambda_2, \lambda - \lambda_2)B^{\frac{1}{q}}(\lambda_1, \lambda - \lambda_1)$$
$$= B(\lambda_2, \lambda - \lambda_2),$$
$$\times B\left(\frac{\lambda - \lambda_2}{p} + \frac{\lambda_1}{q}, \frac{\lambda - \lambda_1}{q} + \frac{\lambda_2}{p}\right)$$
$$= \lim_{p\to 1^-}\lim_{q\to-\infty} B\left(\frac{\lambda - \lambda_2}{p} + \frac{\lambda_1}{q}, \frac{\lambda - \lambda_1}{q} + \frac{\lambda_2}{p}\right)$$
$$= B(\lambda_2, \lambda - \lambda_2) = B^{\frac{1}{p}}(\lambda_2, \lambda - \lambda_2)B^{\frac{1}{q}}(\lambda_1, \lambda - \lambda_1).$$

(ii) $\Rightarrow$ (iii): In view of the expression, (3.44) keeps the form of equality. By the proof of Theorem 3.15, we have $\lambda_1 + \lambda_2 = \lambda$.

(iii) $\Rightarrow$ (i): For $\lambda_1 + \lambda_2 = \lambda$, we have

$$B^{\frac{1}{p}}(\lambda_2, \lambda - \lambda_2) B^{\frac{1}{q}}(\lambda_1, \lambda - \lambda_1)$$

$$= B\left(\frac{\lambda - \lambda_2}{p} + \frac{\lambda_1}{q}, \frac{\lambda - \lambda_1}{q} + \frac{\lambda_2}{p}\right) = B(\lambda_1, \lambda_2).$$

Both of them are independent of p and q.

Hence, (i) $\Leftrightarrow$ (ii) $\Leftrightarrow$ (iii).

(iii) $\Rightarrow$ (iv): If $\lambda_1 + \lambda_2 = \lambda$, then by Theorem 3.14, the constant factor

$$\frac{1}{\lambda} B^{\frac{1}{p}}(\lambda_2, \lambda - \lambda_2) B^{\frac{1}{q}}(\lambda_1, \lambda - \lambda_1) \left(= \frac{1}{\lambda} B(\lambda_1, \lambda_2)\right)$$

in (3.41) is the best possible.

(iv) $\Rightarrow$ (iii): By Theorem 3.15, we have $\lambda_1 + \lambda_2 = \lambda$.

Hence, (iii) $\Leftrightarrow$ (iv).

Therefore, the statements (i), (ii), (iii), and (iv) are equivalent.

This completes the proof of the theorem. $\qquad\qquad\square$

Theorem 3.19. *Inequality* (3.41) *is equivalent to the following reverse Hilbert-type integral inequality involving one derivative function:*

$$J := \left\{\int_0^\infty x^{q(\lambda_1 + a) - a - 1} \left[\int_0^\infty \frac{g(y)}{(x + y)^{\lambda + 1}} dy\right]^q dx\right\}^{\frac{1}{q}}$$

$$> \frac{1}{\lambda} B^{\frac{1}{p}}(\lambda_2, \lambda - \lambda_2) B^{\frac{1}{q}}(\lambda_1, \lambda - \lambda_1)$$

$$\times \left[\int_0^\infty y^{q(1 - \lambda_2) - a - 1} g'^q(y) dy\right]^{\frac{1}{q}}. \tag{3.45}$$

In particular, for $\lambda_1 + \lambda_2 = \lambda$, we reduce (3.45) to the equivalent form of (3.42) as follows:

$$\left\{\int_0^\infty x^{q\lambda_1 - 1} \left[\int_0^\infty \frac{g(y)}{(x + y)^{\lambda + 1}} dy\right]^q dx\right\}^{\frac{1}{q}}$$

$$> \frac{1}{\lambda} B(\lambda_1, \lambda_2) \left[\int_0^\infty y^{q(1 - \lambda_2) - 1} g'^q(y) dy\right]^{\frac{1}{q}}, \tag{3.46}$$

where the constant factor $\frac{1}{\lambda} B(\lambda_1, \lambda_2)$ is the best possible.

Proof. Suppose that (3.45) is valid. By the reverse Hölder inequality, we have

$$I = \int_0^\infty \left(x^{\frac{1}{q}-\lambda_1-\frac{a}{p}} f(x) \right) \left[x^{\frac{-1}{q}+\lambda_1+\frac{a}{p}} \int_0^\infty \frac{g(y)}{(x+y)^{\lambda+1}} dy \right] dx$$

$$\geq \left[\int_0^\infty x^{p(1-\lambda_1)-a-1} f^p(x) dx \right]^{\frac{1}{p}} J. \tag{3.47}$$

Then, by (3.45), we have (3.41).

On the other hand, assuming that (3.41) is valid, we set

$$f(x) := x^{q(\lambda_1+a)-a-1} \left[\int_0^\infty \frac{g(y)}{(x+y)^{\lambda+1}} dy \right]^{q-1}, \quad x \in \mathbf{R}_+.$$

If $J = \infty$, then (3.45) is naturally valid; if $J = 0$, then it is impossible to make (3.45) valid, namely, $J > 0$. Suppose that $0 < J < \infty$. By (3.41), we have

$$\infty > \int_0^\infty x^{p(1-\lambda_1)-a-1} f^p(x) dx = J^q = I$$

$$> \frac{1}{\lambda} B(\lambda_1, \lambda_2) J^{q-1} \left[\int_0^\infty y^{q(1-\lambda_2)-1} g'^q(y) dy \right]^{\frac{1}{q}} > 0,$$

$$J = \left[\int_0^\infty x^{p(1-\lambda_1)-a-1} f^p(x) dx \right]^{\frac{1}{q}}$$

$$> \frac{1}{\lambda} B(\lambda_1, \lambda_2) \left[\int_0^\infty y^{q(1-\lambda_2)-1} g'^q(y) dy \right]^{\frac{1}{q}},$$

namely, (3.45) follows, which is equivalent to (3.41).

The constant factor $\frac{1}{\lambda} B(\lambda_1, \lambda_2)$ is the best possible in (3.46). Otherwise, by (3.47) (for $a = 0$), we would reach a contradiction that the constant factor in (3.42) is not the best possible.

This completes the proof of the theorem. □

Replacing x by $\frac{1}{x}$, then replacing $x^{\lambda-1} f(\frac{1}{x})$ by $f(x)$ in (3.41) and (3.45), we deduce the following corollary.

Corollary 3.20. For $0 < p < 1$, the following reverse Hilbert-type integral inequalities with a nonhomogeneous kernel involving one *derivative function are equivalent:*

$$\int_0^\infty \int_0^\infty \frac{f(x)g(y)}{(1+xy)^{\lambda+1}} dx\, dy$$

$$> \frac{1}{\lambda} B^{\frac{1}{p}}(\lambda_2, \lambda - \lambda_2) B^{\frac{1}{q}}(\lambda_1, \lambda - \lambda_1)$$

$$\times \left[\int_0^\infty x^{p(\lambda_1 - \lambda) + a - 1} f^p(x) dx \right]^{\frac{1}{p}}$$

$$\times \left[\int_0^\infty y^{q(1-\lambda_2) - a - 1} g'^q(y) dy \right]^{\frac{1}{q}}, \tag{3.48}$$

$$\left\{ \int_0^\infty x^{q(\lambda - \lambda_1 - a + 1) + a - 1} \left[\int_0^\infty \frac{g(y)}{(1+xy)^{\lambda+1}} dy \right]^q dx \right\}^{\frac{1}{q}}$$

$$> \frac{1}{\lambda} B^{\frac{1}{p}}(\lambda_2, \lambda - \lambda_2) B^{\frac{1}{q}}(\lambda_1, \lambda - \lambda_1)$$

$$\times \left[\int_0^\infty y^{q(1-\lambda_2) - a - 1} g'^q(y) dy \right]^{\frac{1}{q}}. \tag{3.49}$$

Moreover, $\lambda_1 + \lambda_2 = \lambda$ if and only if the constant factor

$$\frac{1}{\lambda} B^{\frac{1}{p}}(\lambda_2, \lambda - \lambda_2) B^{\frac{1}{q}}(\lambda_1, \lambda - \lambda_1)$$

in (3.48) and (3.49) is the best possible.

For $\lambda_1 + \lambda_2 = \lambda$, we have the following reverse equivalent Hilbert-type inequalities with a best possible constant factor:

$$\int_0^\infty \int_0^\infty \frac{f(x)g(y)}{(1+xy)^{\lambda+1}} dx\, dy$$

$$> \frac{1}{\lambda} B(\lambda_1, \lambda_2) \left(\int_0^\infty x^{-p\lambda_2 - 1} f^p(x) dx \right)^{\frac{1}{p}}$$

$$\times \left[\int_0^\infty y^{q(1-\lambda_2) - 1} g'^q(y) dy \right]^{\frac{1}{q}}, \tag{3.50}$$

$$\left\{ \int_0^\infty x^{q(1+\lambda_2)-1} \left[\int_0^\infty \frac{g(y)}{(1+xy)^{\lambda+1}} dy \right]^q dx \right\}^{\frac{1}{q}}$$
$$> \frac{1}{\lambda} B(\lambda_1, \lambda_2) \left[\int_0^\infty y^{q(1-\lambda_2)-1} g'^q(y) dy \right]^{\frac{1}{q}}. \tag{3.51}$$

Remark 3.21. For

$$\lambda_1 = \frac{\lambda}{r}, \ \lambda_2 = \frac{\lambda}{s} \ \left(r > 1, \frac{1}{r} + \frac{1}{s} = 1 \right)$$

in (3.42), (3.46), (3.48), and (3.49), we have the following two couples of equivalent integral inequalities:

$$\int_0^\infty \int_0^\infty \frac{f(x)g(y)}{(x+y)^{\lambda+1}} dx\, dy > \frac{1}{\lambda} B\left(\frac{\lambda}{r}, \frac{\lambda}{s}\right)$$
$$\times \left[\int_0^\infty x^{p(1-\frac{\lambda}{r})-1} f^p(x) dx \right]^{\frac{1}{p}}$$
$$\times \left[\int_0^\infty y^{q(1-\frac{\lambda}{s})-1} g'^q(y) dy \right]^{\frac{1}{q}}, \tag{3.52}$$

$$\left\{ \int_0^\infty x^{\frac{q\lambda}{r}-1} \left[\int_0^\infty \frac{g(y)}{(x+y)^{\lambda+1}} dy \right]^q dx \right\}^{\frac{1}{q}}$$
$$> \frac{1}{\lambda} B\left(\frac{\lambda}{r}, \frac{\lambda}{s}\right) \left[\int_0^\infty y^{q(1-\frac{\lambda}{s})-1} g'^q(y) dy \right]^{\frac{1}{q}}; \tag{3.53}$$

$$\int_0^\infty \int_0^\infty \frac{f(x)g(y)}{(1+xy)^{\lambda+1}} dx\, dy$$
$$> \frac{1}{\lambda} B\left(\frac{\lambda}{r}, \frac{\lambda}{s}\right) \left(\int_0^\infty x^{-\frac{p\lambda}{s}-1} f^p(x) dx \right)^{\frac{1}{p}}$$
$$\times \left[\int_0^\infty y^{q(1-\frac{\lambda}{s})-1} g'^q(y) dy \right]^{\frac{1}{q}}, \tag{3.54}$$

$$\left\{ \int_0^\infty x^{q(1+\frac{\lambda}{s})-1} \left[\int_0^\infty \frac{g(y)}{(1+xy)^{\lambda+1}} dy \right]^q dx \right\}^{\frac{1}{q}}$$

$$> \frac{1}{\lambda} B\left(\frac{\lambda}{r}, \frac{\lambda}{s}\right) \left[\int_0^\infty y^{q(1-\frac{\lambda}{s})-1} g'^q(y) dy \right]^{\frac{1}{q}}, \qquad (3.55)$$

where the constant factor

$$\frac{1}{\lambda} B\left(\frac{\lambda}{r}, \frac{\lambda}{s}\right)$$

is the best possible.

In particular, for $\lambda = 1$ and $r = s = 2$, we have

$$\int_0^\infty \int_0^\infty \frac{f(x)g(y)}{(x+y)^2} dx\, dy$$

$$> \pi \left(\int_0^\infty x^{\frac{p}{2}-1} f^p(x) dx \right)^{\frac{1}{p}} \left(\int_0^\infty y^{\frac{q}{2}-1} g'^q(y) dy \right)^{\frac{1}{q}}, \qquad (3.56)$$

$$\left\{ \int_0^\infty x^{\frac{q}{2}-1} \left[\int_0^\infty \frac{g(y)\,dy}{(x+y)^2} \right]^q dx \right\}^{\frac{1}{q}} > \pi$$

$$\times \left(\int_0^\infty y^{\frac{q}{2}-1} g'^q(y) dy \right)^{\frac{1}{q}}; \qquad (3.57)$$

$$\int_0^\infty \int_0^\infty \frac{f(x)g(y)}{(1+xy)^2} dx\, dy > \pi \left(\int_0^\infty x^{-\frac{p}{2}-1} f^p(x) dx \right)^{\frac{1}{p}}$$

$$\times \left(\int_0^\infty y^{\frac{q}{2}-1} g'^q(y) dy \right)^{\frac{1}{q}}, \qquad (3.58)$$

$$\left\{ \int_0^\infty x^{\frac{3q}{2}-1} \left[\int_0^\infty \frac{g(y)}{(1+xy)^2} dy \right]^q dx \right\}^{\frac{1}{q}} > \pi$$

$$\times \left(\int_0^\infty y^{\frac{q}{2}-1} g'^q(y) dy \right)^{\frac{1}{q}}. \qquad (3.59)$$

Chapter 4

A New Hilbert-Type Integral Inequality Involving Two Derivative Functions and Others

In this chapter, using weight functions, the idea of introducing parameters and techniques of real analysis, and applying the extended Hardy–Hilbert integral inequality, we prove a new Hilbert-type integral inequality with the homogeneous kernel

$$\frac{1}{(x+y)^{\lambda+2}} \ (\lambda > 0)$$

involving two derivative functions and the beta function. The equivalent statements of the best possible constant factor related to several parameters are considered. In the form of applications, a few particular inequalities and the reverses are obtained. We also consider the case of an inequality involving one derivative function and one upper-limit function.

4.1   Some Lemmas

Hereinafter in Sections 4.1–4.3 of this chapter, we assume that $p > 0$ $(p \neq 1), \frac{1}{p} + \frac{1}{q} = 1, \lambda > 0, \lambda_i \in (0, \lambda) \ (i = 1, 2), a := \lambda - \lambda_1 - \lambda_2,$ and $f(x)$ and $g(y)$ are nonnegative increasing functions in $\mathbf{R}_+$, with

$$f'(x) \cdot g'(y) \geq 0, \quad f(x) = g(y) = o(1) \ (x, y \to 0^+), \text{ and}$$
$$e^{-tx} f(x), e^{-ty} g(y) \in L(\mathbf{R}_+)(t > 0),$$

satisfying

$$0 < \int_0^\infty x^{p(1-\lambda_1)-a-1} f'^p(x) dx < \infty \quad \text{and}$$

$$0 < \int_0^\infty y^{q(1-\lambda_2)-a-1} g'^q(y) dy < \infty.$$

By the definition of the gamma function, for $\lambda, x, y > 0$, the following expression holds (cf. Ref. [123]):

$$\frac{1}{(x+y)^\lambda} = \frac{1}{\Gamma(\lambda)} \int_0^\infty t^{\lambda-1} e^{-(x+y)t} dt. \tag{4.1}$$

Lemma 4.1 (cf. Lemma 3.1). For $t > 0$, we have the following expressions:

$$\int_0^\infty e^{-tx} f(x) dx = \frac{1}{t} \int_0^\infty e^{-tx} f'(x) dx, \tag{4.2}$$

$$\int_0^\infty e^{-ty} g(y) dy = \frac{1}{t} \int_0^\infty e^{-ty} g'(y) dy. \tag{4.3}$$

Lemma 4.2 (cf. Lemma 3.2). Define the following weight functions:

$$\bar{\omega}(\lambda_2, x) := x^{\lambda-\lambda_2} \int_0^\infty \frac{y^{\lambda_2-1}}{(x+y)^\lambda} dy \ (x \in \mathbf{R}_+), \tag{4.4}$$

$$\omega(\lambda_1, y) := y^{\lambda-\lambda_1} \int_0^\infty \frac{x^{\lambda_1-1}}{(x+y)^\lambda} dx \ (y \in \mathbf{R}_+). \tag{4.5}$$

We have the following expressions:

$$\bar{\omega}(\lambda_2, x) = B(\lambda_2, \lambda - \lambda_2) \ (x \in \mathbf{R}_+), \tag{4.6}$$

$$\omega(\lambda_1, y) = B(\lambda_1, \lambda - \lambda_1) \ (y \in \mathbf{R}_+), \tag{4.7}$$

where

$$B(u, v) := \int_0^\infty \frac{t^{u-1}}{(1+t)^{u+v}} dt \quad (u, v > 0)$$

is the beta function.

Lemma 4.3. *For $p > 1$, we have the following extended Hardy–Hilbert integral inequality:*

$$I_1 := \int_0^\infty \int_0^\infty \frac{f'(x)g'(y)}{(x+y)^\lambda} dx\, dy$$

$$< B^{\frac{1}{p}}(\lambda_2, \lambda - \lambda_2) B^{\frac{1}{q}}(\lambda_1, \lambda - \lambda_1)$$

$$\times \left[\int_0^\infty x^{p(1-\lambda_1)-a-1} f'^p(x) dx \right]^{\frac{1}{p}} \left[\int_0^\infty y^{q(1-\lambda_2)-a-1} g'^q(y) dy \right]^{\frac{1}{q}}.$$

$$(4.8)$$

Proof. By Hölder's inequality (cf. Ref. [120]), we obtain

$$I_1 = \int_0^\infty \int_0^\infty \frac{1}{(x+y)^\lambda} \left[\frac{y^{(\lambda_2-1)/p}}{x^{(\lambda_1-1)/q}} f'(x) \right] \left[\frac{x^{(\lambda_1-1)/q}}{y^{(\lambda_2-1)/p}} g'(y) \right] dx\, dy$$

$$\leq \left[\int_0^\infty \varpi(\lambda_2, x) x^{p(1-\lambda_1)-a-1} f'^p(x) dx \right]^{\frac{1}{p}}$$

$$\times \left[\int_0^\infty \omega(\lambda_1, y) y^{q(1-\lambda_2)-a-1} g'^q(y) dy \right]^{\frac{1}{q}}.$$

$$(4.9)$$

If (4.9) retains the form of equality, then there exist constants A and B such that they are not both zero, satisfying

$$A \frac{y^{\lambda_2-1}}{x^{(\lambda_1-1)(p-1)}} f'^p(x) = B \frac{x^{\lambda_1-1}}{y^{(\lambda_2-1)(q-1)}} g'^q(y) \quad \text{a.e. in } \mathbf{R}_+^2.$$

We assume that $A \neq 0$. Then, there exists a $y \in \mathbf{R}_+$ such that

$$x^{p(1-\lambda_1)-a-1} f'^p(x) = \frac{B g'^q(y)}{A y^{q(\lambda_2-1)}} x^{-a-1} \quad \text{a.e. in } \mathbf{R}_+,$$

which contradicts the fact that

$$0 < \int_0^\infty x^{p(1-\lambda_1)-a-1} f'^p(x) dx < \infty$$

since for any $a = \lambda - \lambda_1 - \lambda_2 \in \mathbf{R}$,

$$\int_0^\infty x^{-a-1}dx = \infty.$$

Therefore, by (4.6) and (4.7), we have (4.8).

This completes the proof of the lemma.   □

Remark 4.4. For $0 < p < 1$ ($q < 0$), similarly, by using the reverse Hölder inequality (cf. Ref. [120]), we also derive the reverse of (4.8).

4.2   Main Results and Some Particular Inequalities

Theorem 4.5. *For $p > 1$, we have the following Hilbert-type integral inequality involving two derivative functions:*

$$I := \int_0^\infty \int_0^\infty \frac{f(x)g(y)}{(x+y)^{\lambda+2}}dx\,dy$$

$$< \frac{1}{\lambda(\lambda+1)}B^{\frac{1}{p}}(\lambda_2, \lambda - \lambda_2)B^{\frac{1}{q}}(\lambda_1, \lambda - \lambda_1)$$

$$\times \left[\int_0^\infty x^{p(1-\lambda_1)-a-1}f'^p(x)dx\right]^{\frac{1}{p}}\left[\int_0^\infty y^{q(1-\lambda_2)-a-1}g'^q(y)dy\right]^{\frac{1}{q}}.$$

$$(4.10)$$

In particular, for $\lambda_1 + \lambda_2 = \lambda$ (or $a = 0$), we reduce (4.10) as follows:

$$\int_0^\infty \int_0^\infty \frac{f(x)g(y)}{(x+y)^{\lambda+2}}dx\,dy$$

$$< \frac{B(\lambda_1, \lambda_2)}{\lambda(\lambda+1)}\left[\int_0^\infty x^{p(1-\lambda_1)-1}f'^p(x)dx\right]^{\frac{1}{p}}$$

$$\times \left[\int_0^\infty y^{q(1-\lambda_2)-1}g'^q(y)dy\right]^{\frac{1}{q}},$$

$$(4.11)$$

where the constant factor

$$\frac{1}{\lambda(\lambda+1)}B(\lambda_1, \lambda_2)$$

is the best possible.

Proof. Using Fubini's theorem (cf. Ref. [119]) and (4.1)–(4.3), we obtain that

$$I = \frac{1}{\Gamma(\lambda+2)} \int_0^\infty \int_0^\infty f(x)g(y) \left[\int_0^\infty e^{-(x+y)t} t^{(\lambda+2)-1} dt \right] dx\, dy$$

$$= \frac{1}{\Gamma(\lambda+2)} \int_0^\infty t^{(\lambda+2)-1} \left(\int_0^\infty e^{-xt} f(x) dx \right)$$

$$\times \left(\int_0^\infty e^{-ty} g(y) dy \right) dt$$

$$= \frac{1}{\Gamma(\lambda+2)} \int_0^\infty t^{\lambda-1} \left(\int_0^\infty e^{-xt} f'(x) dx \right)$$

$$\times \left(\int_0^\infty e^{-ty} g'(y) dy \right) dt$$

$$= \frac{1}{\Gamma(\lambda+2)} \int_0^\infty \int_0^\infty f'(x)g'(y) \left[\int_0^\infty e^{-(x+y)t} t^{\lambda-1} dt \right] dx\, dy$$

$$= \frac{\Gamma(\lambda)}{\Gamma(\lambda+2)} \int_0^\infty \int_0^\infty \frac{f'(x)g'(y)}{(x+y)^\lambda} dx\, dy = \frac{1}{\lambda(\lambda+1)} I_1. \qquad (4.12)$$

Then, by (4.8), we derive (4.10).
For $a = 0$ in (4.10), we get (4.11). For any

$$0 < \varepsilon < \min\{p\lambda_1, q\lambda_2\},$$

we set

$$\widetilde{f}(x) := \begin{cases} 0, & 0 < x \le 1, \\ x^{\lambda_1 - \frac{\varepsilon}{p}}, & x > 1; \end{cases} \quad \widetilde{g}(y) := \begin{cases} 0, & 0 < y \le 1, \\ y^{\lambda_2 - \frac{\varepsilon}{q}}, & y > 1. \end{cases}$$

We obtain that

$$\widetilde{f}(x) = \widetilde{g}(y) = o(1)\ (x, y \to 0^+), \quad \widetilde{f}'(x) = 0\ (0 < x < 1),$$

$$\widetilde{g}'(y) = 0\ (0 < y < 1) \quad \text{and}$$

$$\widetilde{f}'(x) = \left(\lambda_1 - \frac{\varepsilon}{p} \right) x^{\lambda_1 - \frac{\varepsilon}{p} - 1}\ (x > 1),$$

$$\widetilde{g}'(y) = \left(\lambda_2 - \frac{\varepsilon}{q} \right) y^{\lambda_2 - \frac{\varepsilon}{q} - 1}\ (y > 1).$$

If there exists a positive constant $M(\leq \frac{1}{\lambda(\lambda+1)}B(\lambda_1, \lambda_2))$ such that (4.11) is valid when replacing

$$\frac{1}{\lambda(\lambda+1)}B(\lambda_1, \lambda_2)$$

by M, then in particular, replacing $f(x), g(y), f'(x)$, and $g'(y)$ by $\widetilde{f}(x), \widetilde{g}(y), \widetilde{f}'(x)$, and $\widetilde{g}'(y)$, respectively, we have

$$\widetilde{I} := \int_0^\infty \int_0^\infty \frac{\widetilde{f}(x)\widetilde{g}(y)}{(x+y)^{\lambda+2}} dx\, dy$$

$$< M \left[\int_0^\infty x^{p(1-\lambda_1)-1} \widetilde{f}'^p(x) dx \right]^{\frac{1}{p}} \left[\int_0^\infty y^{q(1-\lambda_2)-1} \widetilde{g}'^q(y) dy \right]^{\frac{1}{q}}$$

$$= M \left(\lambda_1 - \frac{\varepsilon}{p} \right) \left(\lambda_2 - \frac{\varepsilon}{q} \right) \int_1^\infty x^{-\varepsilon-1} dx$$

$$= \frac{M}{\varepsilon} \left(\lambda_1 - \frac{\varepsilon}{p} \right) \left(\lambda_2 - \frac{\varepsilon}{q} \right).$$

In view of Fubini's theorem (cf. Ref. [119]), setting $u = \frac{y}{x}$, it follows that

$$\widetilde{I} = \int_1^\infty x^{\lambda_1 - \frac{\varepsilon}{p}} \left[\int_1^\infty \frac{y^{\lambda_2 - \frac{\varepsilon}{q}}}{(x+y)^{\lambda+2}} dy \right] dx$$

$$= \int_1^\infty x^{-\varepsilon-1} \left[\int_{\frac{1}{x}}^\infty \frac{u^{\lambda_2 - \frac{\varepsilon}{q}}}{(1+u)^{\lambda+2}} du \right] dx$$

$$= \int_1^\infty x^{-\varepsilon-1} \left[\int_{\frac{1}{x}}^1 \frac{u^{\lambda_2 - \frac{\varepsilon}{q}}}{(1+u)^{\lambda+2}} du \right] dx$$

$$+ \int_1^\infty x^{-\varepsilon-1} \left[\int_1^\infty \frac{u^{\lambda_2 - \frac{\varepsilon}{q}}}{(1+u)^{\lambda+2}} du \right] dx$$

$$= \int_0^1 \left(\int_{\frac{1}{u}}^\infty x^{-\varepsilon-1} dx \right) \frac{u^{\lambda_2 - \frac{\varepsilon}{q}} du}{(1+u)^{\lambda+2}} + \frac{1}{\varepsilon} \int_1^\infty \frac{u^{\lambda_2 - \frac{\varepsilon}{q}} du}{(1+u)^{\lambda+2}}$$

$$= \frac{1}{\varepsilon} \left[\int_0^1 \frac{u^{\lambda_2 + \frac{\varepsilon}{p}}}{(1+u)^{\lambda+2}} du + \int_1^\infty \frac{u^{\lambda_2 - \frac{\varepsilon}{q}}}{(1+u)^{\lambda+2}} du \right].$$

In virtue of the above results, we deduce that

$$\int_0^1 \frac{u^{\lambda_2+\frac{\varepsilon}{p}}}{(1+u)^{\lambda+2}} du + \int_1^\infty \frac{u^{\lambda_2-\frac{\varepsilon}{q}}}{(1+u)^{\lambda+2}} du$$

$$= \varepsilon \tilde{I} < M \left(\lambda_1 - \frac{\varepsilon}{p} \right) \left(\lambda_2 - \frac{\varepsilon}{q} \right).$$

For $\varepsilon \to 0^+$ in the above inequality, in view of the continuity of the beta function, we obtain that

$$\lambda_1 \lambda_2 M \geq \int_0^\infty \frac{u^{(\lambda_2+1)-1}}{(1+u)^{\lambda+2}} du = B(\lambda_1 + 1, \lambda_2 + 1)$$

$$= \frac{\lambda_1 \lambda_2}{\lambda(\lambda+1)} B(\lambda_1, \lambda_2),$$

namely,

$$M \geq \frac{1}{\lambda(\lambda+1)} B(\lambda_1, \lambda_2),$$

which yields that

$$M = \frac{1}{\lambda(\lambda+1)} B(\lambda_1, \lambda_2)$$

is the best possible constant factor in (4.11).

This completes the proof of the theorem. $\qquad\square$

Remark 4.6. We set

$$\widehat{\lambda_1} := \lambda_1 + \frac{a}{p} = \frac{\lambda - \lambda_2}{p} + \frac{\lambda_1}{q} \quad \text{and}$$

$$\widehat{\lambda_2} := \lambda_2 + \frac{a}{q} = \frac{\lambda - \lambda_1}{q} + \frac{\lambda_2}{p}.$$

It follows that $\widehat{\lambda_1} + \widehat{\lambda_2} = \lambda$. We obtain that

$$0 < \widehat{\lambda_1} < \frac{\lambda}{p} + \frac{\lambda}{q} = \lambda \quad \text{and}$$

$$0 < \widehat{\lambda_2} = \lambda - \widehat{\lambda_1} < \lambda, \quad B(\widehat{\lambda_1}, \widehat{\lambda_2}) \in \mathbf{R}_+.$$

So, we rewrite (4.10) as follows:

$$I = \int_0^\infty \int_0^\infty \frac{f(x)g(y)}{(x+y)^{\lambda+2}} dx\, dy$$

$$< \frac{1}{\lambda(\lambda+1)} B^{\frac{1}{p}}(\lambda_2, \lambda - \lambda_2) B^{\frac{1}{q}}(\lambda_1, \lambda - \lambda_1)$$

$$\times \left[\int_0^\infty x^{p(1-\widehat{\lambda}_1)-1} f^p(x) dx \right]^{\frac{1}{p}} \left[\int_0^\infty y^{q(1-\widehat{\lambda}_2)-1} g'^q(y) dy \right]^{\frac{1}{q}}.$$

$$(4.13)$$

Theorem 4.7. *For $p > 1$, if the constant factor*

$$\frac{1}{\lambda(\lambda+1)} B^{\frac{1}{p}}(\lambda_2, \lambda - \lambda_2) B^{\frac{1}{q}}(\lambda_1, \lambda - \lambda_1)$$

in (4.10) *(or* (4.13)*) is the best possible, then $a = 0$, namely,*
$\lambda_1 + \lambda_2 = \lambda$.

Proof. By Hölder's inequality (cf. Ref. [120]), we obtain that

$$B(\widehat{\lambda}_1, \widehat{\lambda}_2) = \int_0^\infty \frac{1}{(1+u)^\lambda} \left(u^{\frac{\lambda-\lambda_2-1}{p}} \right) \left(u^{\frac{\lambda_1-1}{q}} \right) du$$

$$\leq \left[\int_0^\infty \frac{u^{\lambda-\lambda_2-1}}{(1+u)^\lambda} du \right]^{\frac{1}{p}} \left[\int_0^\infty \frac{u^{\lambda_1-1}}{(1+u)^\lambda} du \right]^{\frac{1}{q}}$$

$$= B^{\frac{1}{p}}(\lambda_2, \lambda - \lambda_2) B^{\frac{1}{q}}(\lambda_1, \lambda - \lambda_1). \qquad (4.14)$$

By (4.11) (for $\lambda_i = \widehat{\lambda}_i$ $(i = 1, 2)$), since

$$\frac{1}{\lambda(\lambda+1)} B^{\frac{1}{p}}(\lambda_2, \lambda - \lambda_2) B^{\frac{1}{q}}(\lambda_1, \lambda - \lambda_1)$$

is the best possible constant factor in (4.13), we have

$$\frac{1}{\lambda(\lambda+1)} B^{\frac{1}{p}}(\lambda_2, \lambda - \lambda_2) B^{\frac{1}{q}}(\lambda_1, \lambda - \lambda_1)$$

$$\leq \frac{1}{\lambda(\lambda+1)} B(\widehat{\lambda}_1, \widehat{\lambda}_2)(\in \mathbf{R}_+),$$

namely,

$$B^{\frac{1}{p}}(\lambda_2, \lambda - \lambda_2)B^{\frac{1}{q}}(\lambda_1, \lambda - \lambda_1) \leq B(\widehat{\lambda}_1, \widehat{\lambda}_2).$$

It follows that (4.14) retains the form of equality.

We observe that (4.14) retains the form of equality if and only if there exist constants A and B such that they are not both zero and (cf. Ref. [120]).

$$Au^{\lambda - \lambda_2 - 1} = Bu^{\lambda_1 - 1} \quad \text{a.e. in } \mathbf{R}_+.$$

Assuming that $A \neq 0$, we have

$$u^{\lambda - \lambda_1 - \lambda_2} = \frac{B}{A} \quad \text{a.e. in } \mathbf{R}_+.$$

It follows that $a = \lambda - \lambda_1 - \lambda_2 = 0$, namely, $\lambda_1 + \lambda_2 = \lambda$.

This completes the proof of the theorem. □

Theorem 4.8. *For $p > 1$, the following statements, (i), (ii), (iii), and (iv), are equivalent:*

(i) both $B^{\frac{1}{p}}(\lambda_2, \lambda - \lambda_2)B^{\frac{1}{q}}(\lambda_1, \lambda - \lambda_1)$ and

$$B\left(\frac{\lambda - \lambda_2}{p} + \frac{\lambda_1}{q}, \frac{\lambda - \lambda_1}{q} + \frac{\lambda_1}{p}\right)$$

are independent of p and q;
(ii) the following equality holds:

$$B^{\frac{1}{p}}(\lambda_2, \lambda - \lambda_2)B^{\frac{1}{q}}(\lambda_1, \lambda - \lambda_1)$$
$$= B\left(\frac{\lambda - \lambda_2}{p} + \frac{\lambda_1}{q}, \frac{\lambda - \lambda_1}{q} + \frac{\lambda_1}{p}\right);$$

(iii) $\lambda_1 + \lambda_2 = \lambda$;
(iv) the constant factor

$$\frac{1}{\lambda(\lambda + 1)}B^{\frac{1}{p}}(\lambda_2, \lambda - \lambda_2)B^{\frac{1}{q}}(\lambda_1, \lambda - \lambda_1)$$

is the best possible in (4.10).

Proof. (i) $\Rightarrow$ (ii): In view of the continuity of the beta function, we obtain that

$$B^{\frac{1}{p}}(\lambda_2, \lambda - \lambda_2) B^{\frac{1}{q}}(\lambda_1, \lambda - \lambda_1)$$

$$= \lim_{p \to 1^+} \lim_{q \to \infty} B^{\frac{1}{p}}(\lambda_2, \lambda - \lambda_2) B^{\frac{1}{q}}(\lambda_1, \lambda - \lambda_1) = B(\lambda_2, \lambda - \lambda_2),$$

$$\times B\left(\frac{\lambda - \lambda_2}{p} + \frac{\lambda_1}{q}, \frac{\lambda - \lambda_1}{q} + \frac{\lambda_2}{p}\right)$$

$$= \lim_{p \to 1^+} \lim_{q \to \infty} B\left(\frac{\lambda - \lambda_2}{p} + \frac{\lambda_1}{q}, \frac{\lambda - \lambda_1}{q} + \frac{\lambda_2}{p}\right)$$

$$= B(\lambda_2, \lambda - \lambda_2) = B^{\frac{1}{p}}(\lambda_2, \lambda - \lambda_2) B^{\frac{1}{q}}(\lambda_1, \lambda - \lambda_1).$$

(ii) $\Rightarrow$ (iii): In the assumption, (4.14) retains the form of equality. By the proof of Theorem 4.7, we have $\lambda_1 + \lambda_2 = \lambda$.

(iii) $\Rightarrow$ (iv): If $\lambda_1 + \lambda_2 = \lambda$, then by Theorem 4.5, the constant factor

$$\frac{1}{\lambda(\lambda + 1)} B^{\frac{1}{p}}(\lambda_2, \lambda - \lambda_2) B^{\frac{1}{q}}(\lambda_1, \lambda - \lambda_1)$$

in (4.10) is the best possible.

(iv) $\Rightarrow$ (i): By Theorem 4.7, we have $\lambda_1 + \lambda_2 = \lambda$, and then,

$$B^{\frac{1}{p}}(\lambda_2, \lambda - \lambda_2) B^{\frac{1}{q}}(\lambda_1, \lambda - \lambda_1)$$

$$= B\left(\frac{\lambda - \lambda_2}{p} + \frac{\lambda_1}{q}, \frac{\lambda - \lambda_1}{q} + \frac{\lambda_2}{p}\right) = B(\lambda_1, \lambda_2).$$

Both of them are independent of p and q.

Hence, the statements (i), (ii), (iii), and (iv) are equivalent. This completes the proof of the theorem. $\qquad \square$

Remark 4.9. For $\lambda_1 = \frac{\lambda}{r}, \lambda_2 = \frac{\lambda}{s}$ ($r > 1, \frac{1}{r} + \frac{1}{s} = 1$) in (4.11), we have the following Hilbert-type integral inequality:

$$\int_0^\infty \int_0^\infty \frac{f(x)g(y)}{(x+y)^{\lambda+2}} dx\, dy$$

$$< \frac{B(\frac{\lambda}{r}, \frac{\lambda}{s})}{\lambda(\lambda+1)} \left[\int_0^\infty x^{p(1-\frac{\lambda}{r})-1} f'^p(x) dx \right]^{\frac{1}{p}}$$

$$\times \left[\int_0^\infty y^{q(1-\frac{\lambda}{s})-1} g'^q(y) dy \right]^{\frac{1}{q}}, \tag{4.15}$$

where the constant factor

$$\frac{1}{\lambda(\lambda+1)} B\left(\frac{\lambda}{r}, \frac{\lambda}{s}\right)$$

is the best possible.

In particular:

(i) For $\lambda = 1, r = q$, and $s = p$, we have

$$\int_0^\infty \int_0^\infty \frac{f(x)g(y)}{(x+y)^3} dx\, dy$$

$$< \frac{\pi}{2\sin(\pi/p)} \left(\int_0^\infty f'^p(x) dx \right)^{\frac{1}{p}} \left(\int_0^\infty g'^q(y) dy \right)^{\frac{1}{q}}; \tag{4.16}$$

(ii) for $\lambda = 1, r = p$, and $s = q$, we have the dual form of (4.16) as follows:

$$\int_0^\infty \int_0^\infty \frac{f(x)g(y)}{(x+y)^3} dx\, dy < \frac{\pi}{2\sin(\pi/p)}$$

$$\times \left(\int_0^\infty x^{p-2} f'^p(x) dx \right)^{\frac{1}{p}} \left(\int_0^\infty y^{q-2} g'^q(y) dy \right)^{\frac{1}{q}}; \tag{4.17}$$

(iii) for $p = q = 2$, both (4.16) and (4.17) reduce to

$$\int_0^\infty \int_0^\infty \frac{f(x)g(y)}{(x+y)^3} dx\, dy < \frac{\pi}{2} \left(\int_0^\infty f'^2(x) dx \int_0^\infty g'^2(y) dy \right)^{\frac{1}{2}}. \tag{4.18}$$

4.3   The Reverses

Theorem 4.10. *For $0 < p < 1$, we have the following reverse Hilbert-type integral inequality involving two derivative functions:*

$$I := \int_0^\infty \int_0^\infty \frac{f(x)g(y)}{(x+y)^{\lambda+2}} dx\, dy$$

$$> \frac{1}{\lambda(\lambda+1)} B^{\frac{1}{p}}(\lambda_2, \lambda - \lambda_2) B^{\frac{1}{q}}(\lambda_1, \lambda - \lambda_1)$$

$$\times \left[\int_0^\infty x^{p(1-\lambda_1)-a-1} f'^p(x) dx \right]^{\frac{1}{p}} \left[\int_0^\infty y^{q(1-\lambda_2)-a-1} g'^q(y) dy \right]^{\frac{1}{q}}.$$

$$(4.19)$$

In particular, for $\lambda_1 + \lambda_2 = \lambda$ (or $a = 0$), we reduce (4.19) to the following:

$$\int_0^\infty \int_0^\infty \frac{f(x)g(y)}{(x+y)^{\lambda+2}} dx\, dy$$

$$> \frac{B(\lambda_1, \lambda_2)}{\lambda(\lambda+1)} \left[\int_0^\infty x^{p(1-\lambda_1)-1} f'^p(x) dx \right]^{\frac{1}{p}}$$

$$\times \left[\int_0^\infty y^{q(1-\lambda_2)-1} g'^q(y) dy \right]^{\frac{1}{q}}, \qquad (4.20)$$

where the constant factor

$$\frac{1}{\lambda(\lambda+1)} B(\lambda_1, \lambda_2)$$

is the best possible.

Proof.   By (4.12) and the reverse of (4.8), we derive (4.19).

For $a = 0$ in (4.19), we get (4.20).

For any $0 < \varepsilon < p\lambda_1$, we set $\widetilde{f}(x)$ and $\widetilde{g}(y)$ as in Theorem 4.5. If there exists a positive constant

$$M \left(\geq \frac{1}{\lambda(\lambda+1)} B(\lambda_1, \lambda_2) \right)$$

such that (4.20) is valid when replacing

$$\frac{1}{\lambda(\lambda+1)}B(\lambda_1,\lambda_2)$$

by M, then in particular, replacing $f(x), g(y), f'(x)$, and $g'(y)$ by $\widetilde{f}(x), \widetilde{g}(y), \widetilde{f}'(x)$, and $\widetilde{g}'(y)$, respectively, we have

$$\widetilde{I} := \int_0^\infty \int_0^\infty \frac{\widetilde{f}(x)\widetilde{g}(y)}{(x+y)^{\lambda+2}}\,dx\,dy$$

$$> M\left[\int_0^\infty x^{p(1-\lambda_1)-1}\widetilde{f}'^p(x)dx\right]^{\frac{1}{p}}\left[\int_0^\infty y^{q(1-\lambda_2)-1}\widetilde{g}'^q(y)dy\right]^{\frac{1}{q}}$$

$$= \frac{M}{\varepsilon}\left(\lambda_1 - \frac{\varepsilon}{p}\right)\left(\lambda_2 - \frac{\varepsilon}{q}\right).$$

In view of the proof of Theorem 4.5, by Fubini's theorem, it follows that

$$\widetilde{I} = \frac{1}{\varepsilon}\left[\int_0^1 \frac{u^{\lambda_2+\frac{\varepsilon}{p}}}{(1+u)^{\lambda+2}}du + \int_1^\infty \frac{u^{\lambda_2-\frac{\varepsilon}{q}}}{(1+u)^{\lambda+2}}du\right].$$

So, we obtain

$$\int_0^1 \frac{u^{\lambda_2+\frac{\varepsilon}{p}}}{(1+u)^{\lambda+2}}du + \int_1^\infty \frac{u^{\lambda_2-\frac{\varepsilon}{q}}}{(1+u)^{\lambda+2}}du$$

$$= \varepsilon\widetilde{I} > M\left(\lambda_1 - \frac{\varepsilon}{p}\right)\left(\lambda_2 - \frac{\varepsilon}{q}\right).$$

For $\varepsilon \to 0^+$ in the above inequality, in view of the continuity of the beta function, we deduce

$$\lambda_1\lambda_2 M \leq \int_0^\infty \frac{u^{(\lambda_2+1)-1}du}{(1+u)^{\lambda+2}} = \frac{\lambda_1\lambda_2}{\lambda(\lambda+1)}B(\lambda_1,\lambda_2),$$

namely,

$$M \leq \frac{1}{\lambda(\lambda+1)}B(\lambda_1,\lambda_2).$$

Hence,

$$M = \frac{1}{\lambda(\lambda+1)} B(\lambda_1, \lambda_2)$$

is the best possible constant factor in (4.20).

This completes the proof of the theorem. □

Remark 4.11. We set $\widehat{\lambda}_1$ and $\widehat{\lambda}_2$ as in Remark 4.6. It follows that $\widehat{\lambda}_1 + \widehat{\lambda}_2 = \lambda$. We obtain that for

$$a = \lambda - \lambda_1 - \lambda_2 \in (-p\lambda_1, p(\lambda - \lambda_1)),$$

we have $0 < \widehat{\lambda}_1 < \lambda$. Hence, we still have $0 < \widehat{\lambda}_2 = \lambda - \widehat{\lambda}_1 < \lambda$, and then, $B(\widehat{\lambda}_1, \widehat{\lambda}_2) \in \mathbf{R}_+$. So, we rewrite (4.19) as follows:

$$I = \int_0^\infty \int_0^\infty \frac{f(x)g(y)}{(x+y)^{\lambda+2}} dx\, dy$$
$$> \frac{1}{\lambda(\lambda+1)} B^{\frac{1}{p}}(\lambda_2, \lambda - \lambda_2) B^{\frac{1}{q}}(\lambda_1, \lambda - \lambda_1)$$
$$\times \left[\int_0^\infty x^{p(1-\widehat{\lambda}_1)-1} f^p(x) dx\right]^{\frac{1}{p}} \left[\int_0^\infty y^{q(1-\widehat{\lambda}_2)-1} g'^q(y) dy\right]^{\frac{1}{q}}.$$
$$(4.21)$$

Theorem 4.12. *For $0 < p < 1$, if $a = \lambda - \lambda_1 - \lambda_2 \in (-p\lambda_1, p(\lambda - \lambda_1))$ and the constant factor*

$$\frac{1}{\lambda(\lambda+1)} B^{\frac{1}{p}}(\lambda_2, \lambda - \lambda_2) B^{\frac{1}{q}}(\lambda_1, \lambda - \lambda_1)$$

in (4.19) (or (4.21)) is the best possible, then $a = 0$, namely, $\lambda_1 + \lambda_2 = \lambda$.

Proof. By the reverse Hölder inequality (cf. Ref. [120]), we obtain

$$B(\widehat{\lambda}_1, \widehat{\lambda}_2) = \int_0^\infty \frac{1}{(1+u)^\lambda} \left(u^{\frac{\lambda-\lambda_2-1}{p}} \right) \left(u^{\frac{\lambda_1-1}{q}} \right) du$$

$$\geq \left[\int_0^\infty \frac{u^{\lambda-\lambda_2-1}}{(1+u)^\lambda} du \right]^{\frac{1}{p}} \left[\int_0^\infty \frac{u^{\lambda_1-1}}{(1+u)^\lambda} du \right]^{\frac{1}{q}}$$

$$= B^{\frac{1}{p}}(\lambda_2, \lambda - \lambda_2) B^{\frac{1}{q}}(\lambda_1, \lambda - \lambda_1). \tag{4.22}$$

By (4.20) (for $\lambda_i = \widehat{\lambda}_i$ ($i = 1, 2$)), since

$$\frac{1}{\lambda(\lambda+1)} B^{\frac{1}{p}}(\lambda_2, \lambda - \lambda_2) B^{\frac{1}{q}}(\lambda_1, \lambda - \lambda_1)$$

is the best possible constant factor in (4.21), we have

$$\frac{1}{\lambda(\lambda+1)} B^{\frac{1}{p}}(\lambda_2, \lambda - \lambda_2) B^{\frac{1}{q}}(\lambda_1, \lambda - \lambda_1)$$

$$\geq \frac{1}{\lambda(\lambda+1)} B(\widehat{\lambda}_1, \widehat{\lambda}_2) \ (\in \mathbf{R}_+),$$

namely,

$$B^{\frac{1}{p}}(\lambda_2, \lambda - \lambda_2) B^{\frac{1}{q}}(\lambda_1, \lambda - \lambda_1) \geq B(\widehat{\lambda}_1, \widehat{\lambda}_2).$$

Hence, (4.22) keeps the form of equality. By the proof of Theorem 4.7, it follows that

$$\lambda_1 + \lambda_2 = \lambda.$$

This completes the proof of the theorem. □

Theorem 4.13. *For $0 < p < 1$, the following statements, (i), (ii), (iii), and (iv), are equivalent:*

(i) *both*

$$B^{\frac{1}{p}}(\lambda_2, \lambda - \lambda_2) B^{\frac{1}{q}}(\lambda_1, \lambda - \lambda_1)$$

and

$$B \left(\frac{\lambda - \lambda_2}{p} + \frac{\lambda_1}{q}, \frac{\lambda - \lambda_1}{q} + \frac{\lambda_1}{p} \right)$$

are independent of p and q;

(ii) the following equality holds:

$$B^{\frac{1}{p}}(\lambda_2, \lambda - \lambda_2)B^{\frac{1}{q}}(\lambda_1, \lambda - \lambda_1)$$
$$= B\left(\frac{\lambda - \lambda_2}{p} + \frac{\lambda_1}{q}, \frac{\lambda - \lambda_1}{q} + \frac{\lambda_1}{p}\right);$$

(iii) if

$$a = \lambda - \lambda_1 - \lambda_2 \in (-p\lambda_1, p(\lambda - \lambda_1)),$$

then we have

$$\lambda_1 + \lambda_2 = \lambda;$$

(iv) the constant factor

$$\frac{1}{\lambda(\lambda + 1)}B^{\frac{1}{p}}(\lambda_2, \lambda - \lambda_2)B^{\frac{1}{q}}(\lambda_1, \lambda - \lambda_1)$$

is the best possible in (4.19).

Proof.   (i) $\Rightarrow$ (ii): In view of the continuity of the beta function, we obtain that

$$B^{\frac{1}{p}}(\lambda_2, \lambda - \lambda_2)B^{\frac{1}{q}}(\lambda_1, \lambda - \lambda_1)$$
$$= \lim_{p \to 1^-} \lim_{q \to -\infty} B^{\frac{1}{p}}(\lambda_2, \lambda - \lambda_2)B^{\frac{1}{q}}(\lambda_1, \lambda - \lambda_1) = B(\lambda_2, \lambda - \lambda_2),$$
$$\times B\left(\frac{\lambda - \lambda_2}{p} + \frac{\lambda_1}{q}, \frac{\lambda - \lambda_1}{q} + \frac{\lambda_2}{p}\right)$$
$$= \lim_{p \to 1^-} \lim_{q \to -\infty} B\left(\frac{\lambda - \lambda_2}{p} + \frac{\lambda_1}{q}, \frac{\lambda - \lambda_1}{q} + \frac{\lambda_2}{p}\right)$$
$$= B(\lambda_2, \lambda - \lambda_2) = B^{\frac{1}{p}}(\lambda_2, \lambda - \lambda_2)B^{\frac{1}{q}}(\lambda_1, \lambda - \lambda_1).$$

(ii) $\Rightarrow$ (iii): By the proof of Theorem 4.7, we have $\lambda_1 + \lambda_2 = \lambda$.
(iii) $\Rightarrow$ (i): If $\lambda_1 + \lambda_2 = \lambda$, then

$$B^{\frac{1}{p}}(\lambda_2, \lambda - \lambda_2)B^{\frac{1}{q}}(\lambda_1, \lambda - \lambda_1)$$
$$= B\left(\frac{\lambda - \lambda_2}{p} + \frac{\lambda_1}{q}, \frac{\lambda - \lambda_1}{q} + \frac{\lambda_2}{p}\right) = B(\lambda_1, \lambda_2).$$

Both of them are independent of p and q.

Hence, we have (i) ⇔ (ii) ⇔ (iii).

(iii) ⇒ (iv): By Theorem 4.5, the constant factor

$$\frac{1}{\lambda(\lambda+1)}B^{\frac{1}{p}}(\lambda_2,\lambda-\lambda_2)B^{\frac{1}{q}}(\lambda_1,\lambda-\lambda_1)$$

in (4.19) is the best possible.

(iv) ⇒ (iii): By Theorem 4.7, we have $\lambda_1+\lambda_2=\lambda$.

Hence, (iii) ⇔ (iv), and then the statements (i), (ii), (iii), and (iv) are equivalent.

This completes the proof of the theorem. □

Remark 4.14. For

$$\lambda_1=\frac{\lambda}{r}\quad\text{and}\quad\lambda_2=\frac{\lambda}{s}\left(r>1,\frac{1}{r}+\frac{1}{s}=1\right)$$

in (4.20), we have the following integral inequality:

$$\int_0^\infty\int_0^\infty\frac{f(x)g(y)}{(x+y)^{\lambda+2}}dx\,dy$$
$$>\frac{B(\frac{\lambda}{r},\frac{\lambda}{s})}{\lambda(\lambda+1)}\left[\int_0^\infty x^{p(1-\frac{\lambda}{r})-1}f'^p(x)dx\right]^{\frac{1}{p}}\left[\int_0^\infty y^{q(1-\frac{\lambda}{s})-1}g'^q(y)dy\right]^{\frac{1}{q}},$$
$$(4.23)$$

where the constant factor

$$\frac{1}{\lambda(\lambda+1)}B\left(\frac{\lambda}{r},\frac{\lambda}{s}\right)$$

is the best possible.

In particular, for $r=s=2$, we have

$$\int_0^\infty\int_0^\infty\frac{f(x)g(y)}{(x+y)^{\lambda+2}}dx\,dy$$
$$>\frac{B(\frac{\lambda}{2},\frac{\lambda}{2})}{\lambda(\lambda+1)}\left[\int_0^\infty x^{p(1-\frac{\lambda}{2})-1}f'^p(x)dx\right]^{\frac{1}{p}}$$
$$\times\left[\int_0^\infty y^{q(1-\frac{\lambda}{2})-1}g'^q(y)dy\right]^{\frac{1}{q}}.\qquad(4.24)$$

4.4 The Case of Inequalities Involving One Derivative Function and One Upper-Limit Function

Hereinafter in this section, we assume that $p > 1, \frac{1}{p} + \frac{1}{q} = 1, \lambda > 0$, $\lambda_i \in (0, \lambda)$ $(i = 1, 2)$, $a := \lambda - \lambda_1 - \lambda_2$, $f(x)$ is a nonnegative increasing function in $\mathbf{R}_+$ with

$$f'(x) \geq 0, \ f(x) = o(1) \ (x \to 0^+),$$

and $g(y)$ is a nonnegative measurable function in $\mathbf{R}_+$ such that $e^{-tx} f(x)$ and $e^{-ty} g(y)$ $(t > 0)$ are Lebesgue's integrable functions in $\mathbf{R}_+$,

$$G(y) := \int_0^y g(t)dt,$$

satisfying

$$0 < \int_0^\infty x^{p(1-\lambda_1)-a-1} f'^p(x)dx < \infty \quad \text{and}$$

$$0 < \int_0^\infty y^{-q\lambda_2-a-1} G^q(y)dy < \infty.$$

Lemma 4.15. *For $t > 0$, we have the following equalities:*

$$\int_0^\infty e^{-tx} f(x)dx = \frac{1}{t} \int_0^\infty e^{-tx} f'(x)dx, \tag{4.25}$$

$$\int_0^\infty e^{-ty} g(y)dy = t \int_0^\infty e^{-ty} G(y)dy. \tag{4.26}$$

Proof. We only prove (4.26). Since $G(0) = 0$, we have

$$\int_0^\infty e^{-ty} g(y)dy = \int_0^\infty e^{-ty} dG(y) = e^{-ty} G(y)\big|_0^\infty - \int_0^\infty G(y)de^{-ty}$$

$$= \lim_{y \to \infty} e^{-ty} G(y) + t \int_0^\infty e^{-ty} G(y)dy.$$

In view of the fact that

$$\int_0^\infty e^{-ty} G(y)dy \in [0, \infty),$$

we have

$$\lim_{y \to \infty} e^{-ty} G(y) = 0,$$

and then, (4.26) follows.

This completes the proof of the lemma. □

Lemma 4.16. *Define the following weight functions:*

$$\varpi(\lambda_2, x) := x^{\lambda - \lambda_2} \int_0^\infty \frac{t^{\lambda_2}}{(x+t)^{\lambda+1}} dt \ (x \in \mathbf{R}_+), \qquad (4.27)$$

$$\omega(\lambda_1, y) := y^{\lambda + 1 - \lambda_1} \int_0^\infty \frac{t^{\lambda_1 - 1}}{(t+y)^{\lambda+1}} dt \ (y \in \mathbf{R}_+). \qquad (4.28)$$

We have the following expressions:

$$\varpi(\lambda_2, x) = B(\lambda_2 + 1, \lambda - \lambda_2) \ (x \in \mathbf{R}_+), \qquad (4.29)$$

$$\omega(\lambda_1, y) = B(\lambda_1, \lambda + 1 - \lambda_1) \ (y \in \mathbf{R}_+). \qquad (4.30)$$

Proof.   Setting $u = \frac{t}{x}$, we get

$$\varpi(\lambda_2, x) = x^{\lambda - \lambda_2} \int_0^\infty \frac{(ux)^{\lambda_2}}{(x + ux)^{\lambda+1}} x \, du$$

$$= \int_0^\infty \frac{u^{\lambda_2}}{(1+u)^{\lambda+1}} du = B(\lambda_2 + 1, \lambda - \lambda_2),$$

namely, (4.29) follows. Similarly, we have (4.30).

This completes the proof of the lemma. $\qquad \square$

Lemma 4.17. *We have the following extended Hardy–Hilbert integral inequality:*

$$I_1 := \int_0^\infty \int_0^\infty \frac{f'(x)G(y)}{(x+y)^{\lambda+1}} dx \, dy$$

$$< B^{\frac{1}{p}}(\lambda_2 + 1, \lambda - \lambda_2) B^{\frac{1}{q}}(\lambda_1, \lambda + 1 - \lambda_1)$$

$$\times \left[\int_0^\infty x^{p(1-\lambda_1)-a-1} f'^p(x) dx \right]^{\frac{1}{p}} \left[\int_0^\infty y^{-q\lambda_2 - a - 1} G^q(y) dy \right]^{\frac{1}{q}}.$$
$$(4.31)$$

Proof.   By Hölder's inequality and Fubini's theorem (cf. Ref. [119, 120]), we obtain

$$I_1 = \int_0^\infty \int_0^\infty \frac{1}{(x+y)^{\lambda+1}} \left[\frac{y^{\lambda_2/p}}{x^{(\lambda_1-1)/q}} f'(x) \right] \left[\frac{x^{(\lambda_1-1)/q}}{y^{\lambda_2/p}} G(y) \right] dx \, dy$$

$$\leq \left\{ \int_0^\infty \int_0^\infty \frac{1}{(x+y)^{\lambda+1}} \frac{y^{\lambda_2}}{x^{(\lambda_1-1)(p-1)}} f'^p(x) dy\, dx \right\}^{\frac{1}{p}}$$

$$\times \left\{ \int_0^\infty \int_0^\infty \frac{1}{(x+y)^{\lambda+1}} \frac{x^{\lambda_1-1}}{y^{\lambda_2(q-1)}} G^q(y) dx\, dy \right\}^{\frac{1}{q}}$$

$$= \left[\int_0^\infty \varpi(\lambda_2, x) x^{p(1-\lambda_1)-a-1} f'^p(x) dx \right]^{\frac{1}{p}}$$

$$\times \left[\int_0^\infty \omega(\lambda_1, y) y^{-q\lambda_2-a-1} G^q(y) dy \right]^{\frac{1}{q}}. \tag{4.32}$$

Similarly to how we proved Lemma 4.3, we can show that (4.32) retains the form of equality. Then, by (4.29) and (4.30), we deduce (4.31).

This completes the proof of the lemma. □

Theorem 4.18. *We have the following Hilbert-type integral inequality involving one derivative function and one upper-limit function:*

$$I := \int_0^\infty \int_0^\infty \frac{f(x)g(y)}{(x+y)^{\lambda+1}} dx\, dy$$

$$< B^{\frac{1}{p}}(\lambda_2+1, \lambda-\lambda_2) B^{\frac{1}{q}}(\lambda_1, \lambda+1-\lambda_1)$$

$$\times \left[\int_0^\infty x^{p(1-\lambda_1)-a-1} f'^p(x) dx \right]^{\frac{1}{p}} \left(\int_0^\infty y^{-q\lambda_2-a-1} G^q(y) dy \right)^{\frac{1}{q}}. \tag{4.33}$$

In particular, for $\lambda_1 + \lambda_2 = \lambda$ (or $a = 0$), we reduce (4.33) to the following:

$$\int_0^\infty \int_0^\infty \frac{f(x)g(y)}{(x+y)^{\lambda+1}} dx\, dy$$

$$< \frac{\lambda_2}{\lambda} B(\lambda_1, \lambda_2) \left[\int_0^\infty x^{p(1-\lambda_1)-1} f'^p(x) dx \right]^{\frac{1}{p}}$$

$$\times \left(\int_0^\infty y^{-q\lambda_2-1} G^q(y) dy \right)^{\frac{1}{q}}, \tag{4.34}$$

where the constant factor $\frac{\lambda_2}{\lambda} B(\lambda_1, \lambda_2)$ is the best possible.

Proof. Using Fubini's theorem (cf. Ref. [119]), (4.25) and (4.26), we get

$$
\begin{aligned}
I &= \frac{1}{\Gamma(\lambda+1)} \int_0^\infty \int_0^\infty f(x)g(y) \left[\int_0^\infty e^{-(x+y)t} t^\lambda dt \right] dx\, dy \\
&= \frac{1}{\Gamma(\lambda+1)} \int_0^\infty t^\lambda \left(\int_0^\infty e^{-xt} f(x) dx \right) \left(\int_0^\infty e^{-ty} g(y) dy \right) dt \\
&= \frac{1}{\Gamma(\lambda+1)} \int_0^\infty t^\lambda \left(\int_0^\infty t^{-1} e^{-xt} f'(x) dx \right) \left(\int_0^\infty t e^{-ty} G(y) dy \right) dt \\
&= \frac{1}{\Gamma(\lambda+1)} \int_0^\infty \int_0^\infty f'(x) G(y) \left[\int_0^\infty t^\lambda e^{-(x+y)t} dt \right] dx\, dy = I_1.
\end{aligned}
$$

$$(4.35)$$

Then, by (4.31), we deduce (4.33).
For $a = 0$ in (4.33), we derive (4.34).
For any $0 < \varepsilon < \min\{p\lambda_1, q\lambda_2\}$, we set

$$
\widetilde{f}(x) := \begin{cases} 0, & 0 < x \leq 1, \\ x^{\lambda_1 - \frac{\varepsilon}{p}}, & x > 1; \end{cases} \qquad
\widetilde{g}(y) := \begin{cases} 0, & 0 < y \leq 1, \\ y^{\lambda_2 - \frac{\varepsilon}{q} - 1}, & y > 1. \end{cases}
$$

We obtain that

$$
\widetilde{f}'(x) = \widetilde{G}(y) = 0 \ (0 < x, y \leq 1) \quad \text{and}
$$

$$
\widetilde{f}'(x) = \left(\lambda_1 - \frac{\varepsilon}{p} \right) x^{\lambda_1 - \frac{\varepsilon}{p} - 1} \ (x > 1),
$$

$$
\widetilde{G}(y) = \int_1^y t^{\lambda_2 - \frac{\varepsilon}{q} - 1} dt < \int_0^y t^{\lambda_2 - \frac{\varepsilon}{q} - 1} dt = \frac{y^{\lambda_2 - \frac{\varepsilon}{q}}}{\lambda_2 - \frac{\varepsilon}{q}} (y > 1).
$$

If there exists a positive constant

$$
M \left(\leq \frac{\lambda_2}{\lambda} B(\lambda_1, \lambda_2) \right)
$$

such that (4.34) is valid when replacing

$$
\frac{\lambda_2}{\lambda} B(\lambda_1, \lambda_2)
$$

by M, then in particular, replacing $f(x), g(y), f'(x)$, and $G(y)$ by $\widetilde{f}(x), \widetilde{g}(y), \widetilde{f}'(x)$, and $\widetilde{G}(y)$, respectively, we have

$$\widetilde{I} := \int_0^\infty \int_0^\infty \frac{\widetilde{f}(x)\widetilde{g}(y)}{(x+y)^{\lambda+1}}dx\,dy$$

$$< M\left[\int_0^\infty x^{p(1-\lambda_1)-1}\widetilde{f}'^p(x)dx\right]^{\frac{1}{p}}\left(\int_0^\infty y^{-q\lambda_2-1}\widetilde{G}^q(y)dy\right)^{\frac{1}{q}}$$

$$< \frac{M(\lambda_1-\frac{\varepsilon}{p})}{\lambda_2-\frac{\varepsilon}{q}}\int_1^\infty x^{-\varepsilon-1}dx = \frac{M(\lambda_1-\frac{\varepsilon}{p})}{\varepsilon(\lambda_2-\frac{\varepsilon}{q})}.$$

In view of Fubini's theorem (cf. Ref. [119]), setting $u = \frac{y}{x}$, it follows that

$$\widetilde{I} := \int_1^\infty x^{\lambda_1-\frac{\varepsilon}{p}}\left[\int_1^\infty \frac{y^{\lambda_2-\frac{\varepsilon}{q}-1}}{(x+y)^{\lambda+1}}dy\right]dx$$

$$= \int_1^\infty x^{-\varepsilon-1}\left[\int_{\frac{1}{x}}^\infty \frac{u^{\lambda_2-\frac{\varepsilon}{q}-1}}{(1+u)^{\lambda+1}}du\right]dx$$

$$= \int_1^\infty x^{-\varepsilon-1}\left[\int_{\frac{1}{x}}^1 \frac{u^{\lambda_2-\frac{\varepsilon}{q}-1}}{(1+u)^{\lambda+1}}du\right]dx$$

$$+ \int_1^\infty x^{-\varepsilon-1}\left[\int_1^\infty \frac{u^{\lambda_2-\frac{\varepsilon}{q}-1}}{(1+u)^{\lambda+1}}du\right]dx$$

$$= \int_0^1\left(\int_{\frac{1}{u}}^\infty x^{-\varepsilon-1}dx\right)\frac{u^{\lambda_2-\frac{\varepsilon}{q}-1}du}{(1+u)^{\lambda+1}} + \frac{1}{\varepsilon}\int_1^\infty \frac{u^{\lambda_2-\frac{\varepsilon}{q}-1}du}{(1+u)^{\lambda+1}}$$

$$= \frac{1}{\varepsilon}\left[\int_0^1 \frac{u^{\lambda_2+\frac{\varepsilon}{p}-1}}{(1+u)^{\lambda+1}}du + \int_1^\infty \frac{u^{\lambda_2-\frac{\varepsilon}{q}-1}}{(1+u)^{\lambda+1}}du\right].$$

In virtue of the above results, we obtain

$$\int_0^1 \frac{u^{\lambda_2+\frac{\varepsilon}{p}-1}du}{(1+u)^{\lambda+1}} + \int_1^\infty \frac{u^{\lambda_2-\frac{\varepsilon}{q}-1}du}{(1+u)^{\lambda+1}} = \varepsilon\widetilde{I} < \frac{M(\lambda_1-\frac{\varepsilon}{p})}{\lambda_2-\frac{\varepsilon}{q}}.$$

For $\varepsilon \to 0^+$ in the above inequality, in view of the continuity of the beta function, we obtain that

$$\frac{M\lambda_1}{\lambda_2} \geq \int_0^\infty \frac{u^{\lambda_2-1}du}{(1+u)^{\lambda+1}} = B(\lambda_1+1, \lambda_2) = \frac{\lambda_1}{\lambda}B(\lambda_1, \lambda_2),$$

namely,

$$M \geq \frac{\lambda_2}{\lambda}B(\lambda_1, \lambda_2).$$

Hence,

$$M = \frac{\lambda_2}{\lambda}B(\lambda_1, \lambda_2)$$

is the best possible constant factor in (4.34).

This completes the proof of the theorem. □

Remark 4.19. We set

$$\widehat{\lambda}_1 := \lambda_1 + \frac{a}{p} = \frac{\lambda-\lambda_2}{p} + \frac{\lambda_1}{q} \quad \text{and}$$

$$\widehat{\lambda}_2 := \lambda_2 + \frac{a}{q} = \frac{\lambda-\lambda_1}{q} + \frac{\lambda_2}{p}.$$

It follows that

$$\widehat{\lambda}_1 + \widehat{\lambda}_2 = \lambda.$$

We get that

$$0 < \widehat{\lambda}_1 < \frac{\lambda}{p} + \frac{\lambda}{q} = \lambda \quad \text{and} \quad 0 < \widehat{\lambda}_2 = \lambda - \widehat{\lambda}_1 < \lambda,$$

and then,

$$B(\widehat{\lambda}_1, \widehat{\lambda}_2) \in \mathbf{R}_+.$$

Thus, we rewrite (4.33) as follows:

$$I = \int_0^\infty \int_0^\infty \frac{f(x)g(y)}{(x+y)^{\lambda+1}}dx\,dy$$

$$< B^{\frac{1}{p}}(\lambda_2+1, \lambda-\lambda_2)B^{\frac{1}{q}}(\lambda_1, \lambda+1-\lambda_1)$$

$$\times \left[\int_0^\infty x^{p(1-\widehat{\lambda}_1)-1}f'^p(x)dx\right]^{\frac{1}{p}} \left(\int_0^\infty y^{-q\widehat{\lambda}_2-1}G^q(y)dy\right)^{\frac{1}{q}}.$$

$$(4.36)$$

Theorem 4.20. *If the constant factor*

$$B^{\frac{1}{p}}(\lambda_2 + 1, \lambda - \lambda_2)B^{\frac{1}{q}}(\lambda_1, \lambda + 1 - \lambda_1)$$

in (4.33) *(or* (4.36)*) is the best possible, then* $a = 0$, *namely,* $\lambda_1 + \lambda_2 = \lambda$.

Proof. By Hölder's inequality (cf. Ref. [120]), we obtain

$$B(\widehat{\lambda}_1, \widehat{\lambda}_2 + 1) = \int_0^\infty \frac{1}{(1+u)^{\lambda+1}} \left(u^{\frac{\lambda - \lambda_2 - 1}{p}} \right) \left(u^{\frac{\lambda_1 - 1}{q}} \right) du$$

$$\leq \left[\int_0^\infty \frac{u^{\lambda - \lambda_2 - 1}}{(1+u)^{\lambda+1}} du \right]^{\frac{1}{p}} \left[\int_0^\infty \frac{u^{\lambda_1 - 1}}{(1+u)^{\lambda+1}} du \right]^{\frac{1}{q}}$$

$$= B^{\frac{1}{p}}(\lambda_2 + 1, \lambda - \lambda_2)B^{\frac{1}{q}}(\lambda_1, \lambda + 1 - \lambda_1). \qquad (4.37)$$

By (4.34) (for $\lambda_i = \widehat{\lambda}_i$ $(i = 1, 2)$), since

$$B^{\frac{1}{p}}(\lambda_2 + \overset{\bullet}{1}, \lambda - \lambda_2)B^{\frac{1}{q}}(\lambda_1, \lambda + 1 - \lambda_1)$$

is the best possible constant factor in (4.36), we have

$$B^{\frac{1}{p}}(\lambda_2 + 1, \lambda - \lambda_2)B^{\frac{1}{q}}(\lambda_1, \lambda + 1 - \lambda_1)$$
$$\leq \frac{\widehat{\lambda}_2}{\lambda}B(\widehat{\lambda}_1, \widehat{\lambda}_2) = B(\widehat{\lambda}_1, \widehat{\lambda}_2 + 1)(\in \mathbf{R}_+),$$

which yields that (4.37) retains the form of equality.

We observe that (4.37) keeps the form of equality if and only if there exist constants A and B such that they are not both zero and

$$Au^{\lambda - \lambda_2 - 1} = Bu^{\lambda_1 - 1} \quad \text{a.e. in } \mathbf{R}_+$$

(cf. Ref. [120]). Assuming that $A \neq 0$, we have

$$u^{\lambda - \lambda_1 - \lambda_2} = \frac{B}{A} \quad \text{a.e. in } \mathbf{R}_+.$$

It follows that

$$a = \lambda - \lambda_1 - \lambda_2 = 0,$$

namely,

$$\lambda_1 + \lambda_2 = \lambda.$$

This completes the proof of the theorem. □

Theorem 4.21. *The following statements,* (i), (ii), (iii), *and* (iv), *are equivalent:*

(i) both $B^{\frac{1}{p}}(\lambda_2 + 1, \lambda - \lambda_2)B^{\frac{1}{q}}(\lambda_1, \lambda + 1 - \lambda_1)$ and

$$B\left(\frac{\lambda - \lambda_2}{p} + \frac{\lambda_1}{q}, \frac{\lambda - \lambda_1}{q} + \frac{\lambda_1}{p} + 1\right)$$

are independent of p and q;

(ii) the following equality holds:

$$B^{\frac{1}{p}}(\lambda_2 + 1, \lambda - \lambda_2)B^{\frac{1}{q}}(\lambda_1, \lambda + 1 - \lambda_1)$$
$$= B\left(\frac{\lambda - \lambda_2}{p} + \frac{\lambda_1}{q}, \frac{\lambda - \lambda_1}{q} + \frac{\lambda_1}{p} + 1\right);$$

(iii) $\lambda_1 + \lambda_2 = \lambda$;

(iv) the constant factor

$$B^{\frac{1}{p}}(\lambda_2 + 1, \lambda - \lambda_2)B^{\frac{1}{q}}(\lambda_1, \lambda + 1 - \lambda_1)$$

is the best possible in (4.33).

Proof. (i) $\Rightarrow$ (ii): In view of the continuity of the beta function, we obtain that

$$B^{\frac{1}{p}}(\lambda_2 + 1, \lambda - \lambda_2)B^{\frac{1}{q}}(\lambda_1, \lambda + 1 - \lambda_1)$$

$$= \lim_{p \to 1^+} \lim_{q \to \infty} B^{\frac{1}{p}}(\lambda_2 + 1, \lambda - \lambda_2)B^{\frac{1}{q}}(\lambda_1, \lambda + 1 - \lambda_1)$$

$$= B(\lambda_2 + 1, \lambda - \lambda_2),$$

$$\times B\left(\frac{\lambda - \lambda_2}{p} + \frac{\lambda_1}{q}, \frac{\lambda - \lambda_1}{q} + \frac{\lambda_2}{p} + 1\right)$$

$$= \lim_{p \to 1^+} \lim_{q \to \infty} B\left(\frac{\lambda - \lambda_2}{p} + \frac{\lambda_1}{q}, \frac{\lambda - \lambda_1}{q} + \frac{\lambda_2}{p} + 1\right)$$

$$= B(\lambda_2 + 1, \lambda - \lambda_2) = B^{\frac{1}{p}}(\lambda_2 + 1, \lambda - \lambda_2)B^{\frac{1}{q}}(\lambda_1, \lambda + 1 - \lambda_1).$$

(ii) $\Rightarrow$ (iii): In this condition, (4.37) keeps the form of equality. By the proof of Theorem 4.20, we have $\lambda_1 + \lambda_2 = \lambda$.

(iii) $\Rightarrow$ (iv): If $\lambda_1 + \lambda_2 = \lambda$, then by Theorem 4.5, the constant factor

$$B^{\frac{1}{p}}(\lambda_2 + 1, \lambda - \lambda_2)B^{\frac{1}{q}}(\lambda_1, \lambda + 1 - \lambda_1)\left(=\frac{\lambda_2}{\lambda}B(\lambda_1, \lambda_2)\right)$$

in (4.33) is the best possible.

(iv) $\Rightarrow$ (i): By Theorem 4.20, we have $\lambda_1 + \lambda_2 = \lambda$, and then,

$$B^{\frac{1}{p}}(\lambda_2 + 1, \lambda - \lambda_2)B^{\frac{1}{q}}(\lambda_1, \lambda + 1 - \lambda_1)$$
$$= B\left(\frac{\lambda - \lambda_2}{p} + \frac{\lambda_1}{q}, \frac{\lambda - \lambda_1}{q} + \frac{\lambda_2}{p} + 1\right)$$
$$= B(\lambda_1, \lambda_2 + 1).$$

Both of them are independent of p and q.

Hence, the statements (i), (ii), (iii), and (iv) are equivalent.

This completes the proof of the theorem. □

Remark 4.22. For

$$\lambda_1 = \frac{\lambda}{r} \quad \text{and} \quad \lambda_2 = \frac{\lambda}{s} \quad \left(r > 1, \frac{1}{r} + \frac{1}{s} = 1\right)$$

in (4.34), we have the following Hilbert-type integral inequality:

$$\int_0^\infty \int_0^\infty \frac{f(x)g(y)}{(x+y)^{\lambda+1}}dx\,dy < \frac{1}{s}B\left(\frac{\lambda}{r}, \frac{\lambda}{s}\right)$$
$$\times \left[\int_0^\infty x^{p(1-\frac{\lambda}{r})-1}f'^p(x)dx\right]^{\frac{1}{p}}\left(\int_0^\infty y^{-\frac{q\lambda}{s}-1}G^q(y)dy\right)^{\frac{1}{q}}, \quad (4.38)$$

where the constant factor

$$\frac{1}{s}B\left(\frac{\lambda}{r}, \frac{\lambda}{s}\right)$$

is the best possible.

In particular:

(i) For $\lambda = 1, r = q$, and $s = p$, we have

$$\int_0^\infty \int_0^\infty \frac{f(x)g(y)}{(x+y)^2} dx\, dy$$

$$< \frac{\pi}{p\sin(\pi/p)} \left(\int_0^\infty f'^p(x)dx\right)^{\frac{1}{p}} \left[\int_0^\infty \left(\frac{G(y)}{y}\right)^q dy\right]^{\frac{1}{q}}; \qquad (4.39)$$

(ii) for $\lambda = 1, r = p$, and $s = q$, we have the dual forms of (4.39) as follows:

$$\int_0^\infty \int_0^\infty \frac{f(x)g(y)}{(x+y)^2} dx\, dy$$

$$< \frac{\pi}{q\sin(\pi/p)} \left(\int_0^\infty x^{p-2} f'^p(x)dx\right)^{\frac{1}{p}} \left(\int_0^\infty y^{-2} G^q(y)dy\right)^{\frac{1}{q}}; \qquad (4.40)$$

(iii) for $p = q = 2$, both (4.39) and (4.40) reduce to

$$\int_0^\infty \int_0^\infty \frac{f(x)g(y)}{(x+y)^2} dx\, dy < \frac{\pi}{2} \left[\int_0^\infty f'^2(x)dx \int_0^\infty \left(\frac{G(y)}{y}\right)^2 dy\right]^{\frac{1}{2}}.$$

$$(4.41)$$

Chapter 5

New Hilbert-Type Integral Inequalities Involving Upper-Limit Functions

In this chapter, using weight functions, the idea of introducing parameters and techniques of real analysis, and applying the extended Hardy–Hilbert integral inequality, we prove a new Hilbert-type integral inequality with the homogeneous kernel

$$\frac{1}{(x+y)^\lambda}(\lambda > 0)$$

involving one upper-limit function and the beta function. The equivalent statements of the best possible constant factor related to several parameters are considered. In the form of applications, we obtain the equivalent form, the cases of nonhomogeneous kernel, a few particular inequalities, and operator expressions. We also consider a new case of inequality involving two upper-limit functions.

5.1   Some Lemmas and Main Results

Hereinafter in Sections 5.1–5.3 of this chapter, we assume that $p > 1, \frac{1}{p} + \frac{1}{q} = 1, \lambda > 0, \lambda_i \in (0, \lambda)$ $(i = 1, 2)$, $a := \lambda - \lambda_1 - \lambda_2$, $f(x)$ is a nonnegative measurable function in $\mathbf{R}_+ = (0, \infty)$, and $g(y)$ is a nonnegative measurable function in $\mathbf{R}_+$ such that $e^{-ty}g(y)$ $(t > 0)$

119

is a Lebesgue's integrable function in $\mathbf{R}_+$,

$$G(y) := \int_0^y g(t)dt,$$

satisfying

$$0 < \int_0^\infty x^{p(1-\lambda_1)-a-1} f^p(x)dx < \infty \quad \text{and}$$

$$0 < \int_0^\infty y^{-q\lambda_2-a-1} G^q(y)dy < \infty.$$

Lemma 5.1 (cf. Lemma 4.1). For $t > 0$, we have the following equality:

$$\int_0^\infty e^{-ty} g(y)dy = t \int_0^\infty e^{-ty} G(y)dy. \qquad (5.1)$$

Lemma 5.2 (cf. Lemma 4.16). Define the following weight functions:

$$\varpi(\lambda_2, x) := x^{\lambda-\lambda_2} \int_0^\infty \frac{t^{\lambda_2}}{(x+t)^{\lambda+1}} dt \ (x \in \mathbf{R}_+), \qquad (5.2)$$

$$\omega(\lambda_1, y) := y^{\lambda+1-\lambda_1} \int_0^\infty \frac{t^{\lambda_1-1}}{(t+y)^{\lambda+1}} dt \ (y \in \mathbf{R}_+). \qquad (5.3)$$

We have the following expressions:

$$\varpi(\lambda_2, x) = B(\lambda_2 + 1, \lambda - \lambda_2) \ (x \in \mathbf{R}_+), \qquad (5.4)$$
$$\omega(\lambda_1, y) = B(\lambda_1, \lambda + 1 - \lambda_1) \ (y \in \mathbf{R}_+), \qquad (5.5)$$

where

$$B(u, v) = \int_0^\infty \frac{t^{u-1}}{(1+t)^{u+v}} dt \ (u, v > 0)$$

is the beta function (cf. Ref. [123]).

Lemma 5.3 (cf. Lemma 4.17). We have the following extended Hardy–Hilbert integral inequality:

$$I_1 := \int_0^\infty \int_0^\infty \frac{f(x)G(y)}{(x+y)^{\lambda+1}} dx\, dy$$

$$< B^{\frac{1}{p}}(\lambda_2+1, \lambda-\lambda_2) B^{\frac{1}{q}}(\lambda_1, \lambda+1-\lambda_1)$$

$$\times \left[\int_0^\infty x^{p(1-\lambda_1)-a-1} f^p(x) dx \right]^{\frac{1}{p}} \left(\int_0^\infty y^{-q\lambda_2-a-1} G^q(y) dy \right)^{\frac{1}{q}}.$$

$$(5.6)$$

Theorem 5.4. *We have the following Hilbert-type integral inequality involving one upper-limit function:*

$$I := \int_0^\infty \int_0^\infty \frac{f(x)g(y)}{(x+y)^\lambda} dx\, dy$$

$$< \lambda B^{\frac{1}{p}}(\lambda_2+1, \lambda-\lambda_2) B^{\frac{1}{q}}(\lambda_1, \lambda+1-\lambda_1)$$

$$\times \left[\int_0^\infty x^{p(1-\lambda_1)-a-1} f^p(x) dx \right]^{\frac{1}{p}}$$

$$\times \left(\int_0^\infty y^{-q\lambda_2-a-1} G^q(y) dy \right)^{\frac{1}{q}}.$$

$$(5.7)$$

In particular, for $\lambda_1 + \lambda_2 = \lambda$ (or $a = 0$), we reduce (5.7) to the following:

$$\int_0^\infty \int_0^\infty \frac{f(x)g(y)}{(x+y)^\lambda} dx\, dy$$

$$< \lambda_2 B(\lambda_1, \lambda_2) \left[\int_0^\infty x^{p(1-\lambda_1)-1} f^p(x) dx \right]^{\frac{1}{p}}$$

$$\times \left(\int_0^\infty y^{-q\lambda_2-1} G^q(y) dy \right)^{\frac{1}{q}},$$

$$(5.8)$$

where the constant factor $\lambda_2 B(\lambda_1, \lambda_2)$ is the best possible.

Proof.   Using Fubini's theorem (cf. Ref. [119]) and (5.1), we derive that

$$
\begin{aligned}
I &= \frac{1}{\Gamma(\lambda)} \int_0^\infty \int_0^\infty f(x)g(y) \left[\int_0^\infty e^{-(x+y)t} t^{\lambda-1} dt \right] dx\, dy \\
&= \frac{1}{\Gamma(\lambda)} \int_0^\infty t^{\lambda-1} \left(\int_0^\infty e^{-xt} f(x) dx \right) \left(\int_0^\infty e^{-ty} g(y) dy \right) dt \\
&= \frac{1}{\Gamma(\lambda)} \int_0^\infty t^{\lambda-1} \left(\int_0^\infty e^{-xt} f(x) dx \right) \left(\int_0^\infty t e^{-ty} G(y) dy \right) dt \\
&= \frac{1}{\Gamma(\lambda)} \int_0^\infty \int_0^\infty f(x)G(y) \left[\int_0^\infty e^{-(x+y)t} t^{(\lambda+1)-1} dt \right] dx\, dy \\
&= \frac{\Gamma(\lambda+1)}{\Gamma(\lambda)} \int_0^\infty \int_0^\infty \frac{f(x)G(y)}{(x+y)^{\lambda+1}} dx\, dy = \lambda I_1.
\end{aligned}
\tag{5.9}
$$

Then, by (5.6), we have (5.7).

For $a = 0$ in (5.7), we have (5.8).

For any $0 < \varepsilon < q\lambda_2$, we set

$$
\tilde{f}(x) := \begin{cases} 0, & 0 < x \le 1, \\ x^{\lambda_1 - \frac{\varepsilon}{p} - 1}, & x > 1; \end{cases} \qquad
\tilde{g}(y) := \begin{cases} 0, & 0 < y \le 1, \\ y^{\lambda_2 - \frac{\varepsilon}{q} - 1}, & y > 1. \end{cases}
$$

We obtain that

$$
e^{-ty}\tilde{g}(y) \in L(\mathbf{R}_+), \quad \tilde{G}(y) = 0 \quad (0 < y \le 1),
$$

and

$$
\begin{aligned}
\tilde{G}(y) &= \int_1^y t^{\lambda_2 - \frac{\varepsilon}{q} - 1} dt < \int_0^y t^{\lambda_2 - \frac{\varepsilon}{q} - 1} dt \\
&= \frac{y^{\lambda_2 - \frac{\varepsilon}{q}}}{\lambda_2 - \frac{\varepsilon}{q}} (y > 1).
\end{aligned}
$$

If there exists a positive constant $M(\le \lambda_2 B(\lambda_1, \lambda_2))$ such that (5.8) is valid when replacing $\lambda_2 B(\lambda_1, \lambda_2)$ by M, then in particular, replacing $f(x), g(y),$ and $G(y)$ by $\tilde{f}(x), \tilde{g}(y),$ and $\tilde{G}(y)$, respectively, we have

$$
\widetilde{I} := \int_0^\infty \int_0^\infty \frac{\widetilde{f}(x)\widetilde{g}(y)}{(x+y)^\lambda} dx\, dy
$$

$$
< M \left[\int_0^\infty x^{p(1-\lambda_1)-1} \widetilde{f}^p(x) dx \right]^{\frac{1}{p}} \left(\int_0^\infty y^{-q\lambda_2-1} \widetilde{G}^q(y) dy \right)^{\frac{1}{q}}
$$

$$
< \frac{M}{\lambda_2 - \frac{\varepsilon}{q}} \left(\int_1^\infty x^{-\varepsilon-1} dx \right)^{\frac{1}{p}} \left(\int_1^\infty y^{-\varepsilon-1} dy \right)^{\frac{1}{q}}
$$

$$
= \frac{M}{\varepsilon(\lambda_2 - \frac{\varepsilon}{q})}.
$$

In view of Fubini's theorem (cf. Ref. [119]), setting $u = \frac{y}{x}$, it follows that

$$
\widetilde{I} := \int_1^\infty x^{\lambda_1 - \frac{\varepsilon}{p} - 1} \left[\int_1^\infty \frac{y^{\lambda_2 - \frac{\varepsilon}{q} - 1}}{(x+y)^\lambda} dy \right] dx
$$

$$
= \int_1^\infty x^{-\varepsilon-1} \left[\int_{\frac{1}{x}}^\infty \frac{u^{\lambda_2 - \frac{\varepsilon}{q} - 1}}{(1+u)^\lambda} du \right] dx
$$

$$
= \int_1^\infty x^{-\varepsilon-1} \left[\int_{\frac{1}{x}}^1 \frac{u^{\lambda_2 - \frac{\varepsilon}{q} - 1}}{(1+u)^\lambda} du \right] dx
$$

$$
+ \int_1^\infty x^{-\varepsilon-1} \left[\int_1^\infty \frac{u^{\lambda_2 - \frac{\varepsilon}{q} - 1}}{(1+u)^\lambda} du \right] dx
$$

$$
= \int_0^1 \left(\int_{\frac{1}{u}}^\infty x^{-\varepsilon-1} dx \right) \frac{u^{\lambda_2 - \frac{\varepsilon}{q} - 1} du}{(1+u)^\lambda} + \frac{1}{\varepsilon} \int_1^\infty \frac{u^{\lambda_2 - \frac{\varepsilon}{q} - 1} du}{(1+u)^\lambda}
$$

$$
= \frac{1}{\varepsilon} \left[\int_0^1 \frac{u^{\lambda_2 + \frac{\varepsilon}{p} - 1}}{(1+u)^\lambda} du + \int_1^\infty \frac{u^{\lambda_2 - \frac{\varepsilon}{q} - 1}}{(1+u)^\lambda} du \right].
$$

In virtue of the above results, we obtain

$$
\int_0^1 \frac{u^{\lambda_2 + \frac{\varepsilon}{p} - 1}}{(1+u)^\lambda} du + \int_1^\infty \frac{u^{\lambda_2 - \frac{\varepsilon}{q} - 1}}{(1+u)^\lambda} du = \varepsilon \widetilde{I} < \frac{M}{\lambda_2 - \frac{\varepsilon}{q}}.
$$

For $\varepsilon \to 0^+$ in the above inequality, in view of the continuity of the beta function, we get that

$$\frac{M}{\lambda_2} \geq \int_0^\infty \frac{u^{\lambda_2-1}}{(1+u)^\lambda} du = B(\lambda_1, \lambda_2),$$

namely, $M \geq \lambda_2 B(\lambda_1, \lambda_2)$. Hence,

$$M = \lambda_2 B(\lambda_1, \lambda_2)$$

is the best possible constant factor in (5.8).

This completes the proof of the theorem.   □

Remark 5.5. We set

$$\widehat{\lambda}_1 := \lambda_1 + \frac{a}{p} = \frac{\lambda - \lambda_2}{p} + \frac{\lambda_1}{q} \quad \text{and}$$

$$\widehat{\lambda}_2 := \lambda_2 + \frac{a}{q} = \frac{\lambda - \lambda_1}{q} + \frac{\lambda_2}{p}.$$

It follows that $\widehat{\lambda}_1 + \widehat{\lambda}_2 = \lambda$. We obtain that

$$0 < \widehat{\lambda}_1 < \frac{\lambda}{p} + \frac{\lambda}{q} = \lambda \quad \text{and} \quad 0 < \widehat{\lambda}_2 = \lambda - \widehat{\lambda}_1 < \lambda,$$

and then,

$$B(\widehat{\lambda}_1, \widehat{\lambda}_2) \in \mathbf{R}_+.$$

Therefore, we rewrite (5.7) as follows:

$$I = \int_0^\infty \int_0^\infty \frac{f(x)g(y)}{(x+y)^\lambda} dx\, dy$$

$$< \lambda B^{\frac{1}{p}}(\lambda_2+1, \lambda-\lambda_2) B^{\frac{1}{q}}(\lambda_1, \lambda+1-\lambda_1)$$

$$\times \left[\int_0^\infty x^{p(1-\widehat{\lambda}_1)-1} f^p(x) dx \right]^{\frac{1}{p}}$$

$$\times \left(\int_0^\infty y^{-q\widehat{\lambda}_2-1} G^q(y) dy \right)^{\frac{1}{q}}. \tag{5.10}$$

Theorem 5.6. *If the constant factor*

$$\lambda B^{\frac{1}{p}}(\lambda_2 + 1, \lambda - \lambda_2) B^{\frac{1}{q}}(\lambda_1, \lambda + 1 - \lambda_1)$$

in (5.7) *(or* (5.10)*) is the best possible, then* $a = 0$, *namely,* $\lambda_1 + \lambda_2 = \lambda$.

Proof. By Hölder's inequality (cf. Ref. [120]), we obtain

$$B(\widehat{\lambda}_1, \widehat{\lambda}_2 + 1)$$

$$= \int_0^\infty \frac{u^{\widehat{\lambda}_1 - 1} du}{(1 + u)^{\lambda + 1}} = \int_0^\infty \frac{u^{\frac{\lambda - \lambda_2}{p} + \frac{\lambda_1}{q} - 1}}{(1 + u)^{\lambda + 1}} du$$

$$= \int_0^\infty \frac{1}{(1 + u)^{\lambda + 1}} \left(u^{\frac{\lambda - \lambda_2 - 1}{p}} \right) \left(u^{\frac{\lambda_1 - 1}{q}} \right) du$$

$$\leq \left[\int_0^\infty \frac{u^{\lambda - \lambda_2 - 1}}{(1 + u)^{\lambda + 1}} du \right]^{\frac{1}{p}} \left[\int_0^\infty \frac{u^{\lambda_1 - 1}}{(1 + u)^{\lambda + 1}} du \right]^{\frac{1}{q}}$$

$$= B^{\frac{1}{p}}(\lambda_2 + 1, \lambda - \lambda_2) B^{\frac{1}{q}}(\lambda_1, \lambda + 1 - \lambda_1). \tag{5.11}$$

By (5.8) (for $\lambda_i = \widehat{\lambda}_i$ $(i = 1, 2)$), since

$$\lambda B^{\frac{1}{p}}(\lambda_2 + 1, \lambda - \lambda_2) B^{\frac{1}{q}}(\lambda_1, \lambda + 1 - \lambda_1)$$

is the best possible constant factor in (5.10), we have

$$\lambda B^{\frac{1}{p}}(\lambda_2 + 1, \lambda - \lambda_2) B^{\frac{1}{q}}(\lambda_1, \lambda + 1 - \lambda_1)$$
$$\leq \widehat{\lambda}_2 B(\widehat{\lambda}_1, \widehat{\lambda}_2) = \lambda B(\widehat{\lambda}_1, \widehat{\lambda}_2 + 1)(\in \mathbf{R}_+),$$

namely,

$$B^{\frac{1}{p}}(\lambda_2 + 1, \lambda - \lambda_2) B^{\frac{1}{q}}(\lambda_1, \lambda + 1 - \lambda_1) \leq B(\widehat{\lambda}_1, \widehat{\lambda}_2 + 1).$$

It follows that (5.11) retains the form of equality.

We observe that (5.11) retains the form of equality if and only if there exist constants A and B such that they are not both zero and (cf. Ref. [120])

$$Au^{\lambda - \lambda_2 - 1} = Bu^{\lambda_1 - 1} \quad \text{a.e. in } \mathbf{R}_+.$$

Assuming that $A \neq 0$, we have

$$u^{\lambda - \lambda_1 - \lambda_2} = \frac{B}{A} \quad \text{a.e. in } \mathbf{R}_+.$$

It follows that $a = \lambda - \lambda_1 - \lambda_2 = 0$, namely, $\lambda_1 + \lambda_2 = \lambda$.

This completes the proof of the theorem. $\qquad\qquad\qquad\qquad\square$

Theorem 5.7. *The following statements, (i), (ii), (iii), and (iv), are equivalent:*

(i) *both* $B^{\frac{1}{p}}(\lambda_2 + 1, \lambda - \lambda_2) B^{\frac{1}{q}}(\lambda_1, \lambda + 1 - \lambda_1)$ *and*

$$B\left(\frac{\lambda - \lambda_2}{p} + \frac{\lambda_1}{q}, \frac{\lambda - \lambda_1}{q} + \frac{\lambda_1}{p} + 1\right)$$

are independent of p and q;

(ii) *the following equality holds:*

$$B^{\frac{1}{p}}(\lambda_2 + 1, \lambda - \lambda_2) B^{\frac{1}{q}}(\lambda_1, \lambda + 1 - \lambda_1)$$
$$= B\left(\frac{\lambda - \lambda_2}{p} + \frac{\lambda_1}{q}, \frac{\lambda - \lambda_1}{q} + \frac{\lambda_1}{p} + 1\right);$$

(iii) $\lambda_1 + \lambda_2 = \lambda$;

(iv) *the constant factor*

$$\lambda B^{\frac{1}{p}}(\lambda_2 + 1, \lambda - \lambda_2) B^{\frac{1}{q}}(\lambda_1, \lambda + 1 - \lambda_1)$$

is the best possible in (5.7).

Proof. (i) $\Rightarrow$ (ii): In view of the continuity of the beta function, we get that

$$B^{\frac{1}{p}}(\lambda_2 + 1, \lambda - \lambda_2) B^{\frac{1}{q}}(\lambda_1, \lambda + 1 - \lambda_1)$$
$$= \lim_{p \to 1+} \lim_{q \to \infty} B^{\frac{1}{p}}(\lambda_2 + 1, \lambda - \lambda_2) B^{\frac{1}{q}}(\lambda_1, \lambda + 1 - \lambda_1)$$
$$= B(\lambda_2 + 1, \lambda - \lambda_2),$$

$$\times B\left(\frac{\lambda - \lambda_2}{p} + \frac{\lambda_1}{q}, \frac{\lambda - \lambda_1}{q} + \frac{\lambda_2}{p} + 1\right)$$

$$= \lim_{p \to 1+} \lim_{q \to \infty} B\left(\frac{\lambda - \lambda_2}{p} + \frac{\lambda_1}{q}, \frac{\lambda - \lambda_1}{q} + \frac{\lambda_2}{p} + 1\right)$$

$$= B(\lambda_2 + 1, \lambda - \lambda_2)$$

$$= B^{\frac{1}{p}}(\lambda_2 + 1, \lambda - \lambda_2)B^{\frac{1}{q}}(\lambda_1, \lambda + 1 - \lambda_1).$$

(ii) $\Rightarrow$ (iii): In this condition, (5.11) retains the form of equality. By the proof of Theorem 5.6, we have $\lambda_1 + \lambda_2 = \lambda$.

(iii) $\Rightarrow$ (iv): If $\lambda_1 + \lambda_2 = \lambda$, then by Theorem 5.4, the constant factor

$$\lambda B^{\frac{1}{p}}(\lambda_2 + 1, \lambda - \lambda_2)B^{\frac{1}{q}}(\lambda_1, \lambda + 1 - \lambda_1)(= \lambda B(\lambda_1, \lambda_2))$$

in (5.7) is the best possible.

(iv) $\Rightarrow$ (i): By Theorem 5.6, we have $\lambda_1 + \lambda_2 = \lambda$, and then,

$$B^{\frac{1}{p}}(\lambda_2 + 1, \lambda - \lambda_2)B^{\frac{1}{q}}(\lambda_1, \lambda + 1 - \lambda_1)$$

$$= B\left(\frac{\lambda - \lambda_2}{p} + \frac{\lambda_1}{q}, \frac{\lambda - \lambda_1}{q} + \frac{\lambda_2}{p} + 1\right)$$

$$= B(\lambda_1, \lambda_2 + 1).$$

Both of them are independent of p and q.

Hence, the statements (i), (ii), (iii), and (iv) are equivalent.

This completes the proof of the theorem. $\qquad\square$

5.2 Equivalent Form and Some Particular Inequalities

Theorem 5.8. *Inequality* (5.7) *is equivalent to the following inequality involving one upper-limit function:*

$$J := \left\{\int_0^\infty x^{q(\lambda - \lambda_2) - a - 1}\left[\int_0^\infty \frac{g(y)}{(x + y)^\lambda}dy\right]^q dx\right\}^{\frac{1}{q}}$$

$$< \lambda B^{\frac{1}{p}}(\lambda_2 + 1, \lambda - \lambda_2)B^{\frac{1}{q}}(\lambda_1, \lambda + 1 - \lambda_1)$$

$$\times \left(\int_0^\infty y^{-q\lambda_2 - a - 1}G^q(y)dy\right)^{\frac{1}{q}}. \tag{5.12}$$

In particular, for $\lambda_1 + \lambda_2 = \lambda$, we reduce (5.12) to the equivalent form of (5.8) as follows:

$$\left\{ \int_0^\infty x^{q\lambda_1 - 1} \left[\int_0^\infty \frac{g(y)}{(x+y)^\lambda} dy \right]^q dx \right\}^{\frac{1}{q}}$$

$$< \lambda_2 B(\lambda_1, \lambda_2) \left(\int_0^\infty y^{-q\lambda_2 - 1} G^q(y) dy \right)^{\frac{1}{q}}, \qquad (5.13)$$

where the constant factor $\lambda_2 B(\lambda_1, \lambda_2)$ is the best possible.

Proof.   Suppose that (5.12) is valid. By Hölder's inequality, we have

$$I = \int_0^\infty \left[x^{\frac{1}{q} - \lambda_1 - \frac{a}{p}} f(x) \right] \left[x^{\frac{-1}{q} + \lambda_1 + \frac{a}{p}} \int_0^\infty \frac{g(y)}{(x+y)^\lambda} dy \right] dx$$

$$\leq \left[\int_0^\infty x^{p(1-\lambda_1) - a - 1} f^p(x) dx \right]^{\frac{1}{p}} J. \qquad (5.14)$$

Then, by (5.12), we have (5.7).
On the other hand, assuming that (5.7) is valid, we set

$$f(x) := x^{q(\lambda - \lambda_2) - a - 1} \left[\int_0^\infty \frac{g(y)}{(x+y)^\lambda} dy \right]^{q-1}, \qquad x \in \mathbf{R}_+.$$

If $J = 0$, then (5.12) is naturally valid; if $J = \infty$, then it is impossible to make (5.12) valid, namely, $J < \infty$. Suppose that $0 < J < \infty$. By (5.7), we have

$$0 < \int_0^\infty x^{p(1-\lambda_1) - a - 1} f^p(x) dx = J^q = I$$

$$< \lambda B^{\frac{1}{p}}(\lambda_2 + 1, \lambda - \lambda_2) B^{\frac{1}{q}}(\lambda_1, \lambda + 1 - \lambda_1)$$

$$\times J^{q-1} \left(\int_0^\infty y^{-q\lambda_2 - a - 1} G^q(y) dy \right)^{\frac{1}{q}} < \infty,$$

$$J = \left[\int_0^\infty x^{p(1-\lambda_1)-a-1} f^p(x) dx \right]^{\frac{1}{q}}$$

$$< \lambda B^{\frac{1}{p}}(\lambda_2 + 1, \lambda - \lambda_2) B^{\frac{1}{q}}(\lambda_1, \lambda + 1 - \lambda_1)$$

$$\times \left(\int_0^\infty y^{-q\lambda_2-a-1} G^q(y) dy \right)^{\frac{1}{q}},$$

namely, (5.12) follows, which is equivalent to (5.7).

The constant factor $\lambda_2 B(\lambda_1, \lambda_2)$ is the best possible in (5.13). Otherwise, by (5.14) (for $a = 0$), we would reach a contradiction that the constant factor in (5.13) is not the best possible.

This completes the proof of the theorem. □

Replacing x by $\frac{1}{x}$, then replacing $x^{\lambda-1} f(\frac{1}{x})$ by $f(x)$ in (5.7) and (5.12), we deduce the following corollary.

Corollary 5.9. *The following Hilbert-type integral inequalities with a nonhomogeneous kernel involving one upper-limit function are equivalent:*

$$\int_0^\infty \int_0^\infty \frac{f(x)g(y)}{(1+xy)^\lambda} dx\, dy$$

$$< \lambda B^{\frac{1}{p}}(\lambda_2 + 1, \lambda - \lambda_2) B^{\frac{1}{q}}(\lambda_1, \lambda + 1 - \lambda_1)$$

$$\times \left[\int_0^\infty x^{p(\lambda_1-\lambda)+a-1} f^p(x) dx \right]^{\frac{1}{p}}$$

$$\times \left(\int_0^\infty y^{-q\lambda_2-a-1} G^q(y) dy \right)^{\frac{1}{q}}, \tag{5.15}$$

$$\left\{ \int_0^\infty x^{q\lambda_2 hy+a-1} \left[\int_0^\infty \frac{g(y)}{(1+xy)^\lambda} dy \right]^q dx \right\}^{\frac{1}{q}}$$

$$< \lambda B^{\frac{1}{p}}(\lambda_2 + 1, \lambda - \lambda_2) B^{\frac{1}{q}}(\lambda_1, \lambda + 1 - \lambda_1)$$

$$\times \left(\int_0^\infty y^{-q\lambda_2-a-1} G^q(y) dy \right)^{\frac{1}{q}}. \tag{5.16}$$

Moreover, $\lambda_1 + \lambda_2 = \lambda$ if and only if the constant factor

$$\lambda B^{\frac{1}{p}}(\lambda_2 + 1, \lambda - \lambda_2) B^{\frac{1}{q}}(\lambda_1, \lambda + 1 - \lambda_1)$$

in (5.15) and (5.16) is the best possible.

For $\lambda_1 + \lambda_2 = \lambda$, we have the following equivalent inequalities with the best possible constant factor:

$$\int_0^\infty \int_0^\infty \frac{f(x)g(y)}{(1+xy)^\lambda} dx\, dy < \lambda_2 B(\lambda_1, \lambda_2)$$

$$\times \left[\int_0^\infty x^{p(1-\lambda_2)-1} f^p(x) dx \right]^{\frac{1}{p}} \left(\int_0^\infty y^{-q\lambda_2-1} G^q(y) dy \right)^{\frac{1}{q}},$$

$$(5.17)$$

$$\left\{ \int_0^\infty x^{q\lambda_2-1} \left[\int_0^\infty \frac{g(y)}{(1+xy)^\lambda} dy \right]^q dx \right\}^{\frac{1}{q}}$$

$$< \lambda_2 B(\lambda_1, \lambda_2) \left(\int_0^\infty y^{-q\lambda_2-1} G^q(y) dy \right)^{\frac{1}{q}}.$$

$$(5.18)$$

Remark 5.10. For

$$\lambda_1 = \frac{\lambda}{r} \quad \text{and} \quad \lambda_2 = \frac{\lambda}{s} \quad \left(r > 1, \frac{1}{r} + \frac{1}{s} = 1 \right)$$

in (5.8), (5.13), (5.17), and (5.18), we have the following two couples of equivalent integral inequalities:

$$\int_0^\infty \int_0^\infty \frac{f(x)g(y)}{(x+y)^\lambda} dx\, dy < \frac{\lambda}{s} B\left(\frac{\lambda}{r}, \frac{\lambda}{s} \right)$$

$$\times \left[\int_0^\infty x^{p(1-\frac{\lambda}{r})-1} f^p(x) dx \right]^{\frac{1}{p}} \left(\int_0^\infty y^{-\frac{q\lambda}{s}-1} G^q(y) dy \right)^{\frac{1}{q}}, \quad (5.19)$$

$$\left\{ \int_0^\infty x^{\frac{q\lambda}{r}-1} \left[\int_0^\infty \frac{g(y)}{(x+y)^\lambda} dy \right]^q dx \right\}^{\frac{1}{q}}$$

$$< \frac{\lambda}{s} B\left(\frac{\lambda}{r}, \frac{\lambda}{s} \right) \left(\int_0^\infty y^{-\frac{q\lambda}{s}-1} G^q(y) dy \right)^{\frac{1}{q}}; \quad (5.20)$$

$$\int_0^\infty \int_0^\infty \frac{f(x)g(y)}{(1+xy)^\lambda} dx\, dy < \frac{\lambda}{s} B\left(\frac{\lambda}{r}, \frac{\lambda}{s}\right)$$

$$\times \left[\int_0^\infty x^{p(1-\frac{\lambda}{s})-1} f^p(x) dx\right]^{\frac{1}{p}} \left(\int_0^\infty y^{-\frac{q\lambda}{s}-1} G^q(y) dy\right)^{\frac{1}{q}}, \quad (5.21)$$

$$\left\{\int_0^\infty x^{\frac{q\lambda}{s}-1} \left[\int_0^\infty \frac{g(y)}{(1+xy)^\lambda} dy\right]^q dx\right\}^{\frac{1}{q}}$$

$$< \frac{\lambda}{s} B\left(\frac{\lambda}{r}, \frac{\lambda}{s}\right) \left(\int_0^\infty y^{-\frac{q\lambda}{s}-1} G^q(y) dy\right)^{\frac{1}{q}}, \quad (5.22)$$

where the constant factor

$$\frac{\lambda}{s} B\left(\frac{\lambda}{r}, \frac{\lambda}{s}\right)$$

is the best possible.

In particular:

(i) for $\lambda = 1, r = q$, and $s = p$, we have

$$\int_0^\infty \int_0^\infty \frac{f(x)g(y)}{x+y} dx\, dy$$

$$< \frac{\pi}{p\sin(\pi/p)} \left(\int_0^\infty f^p(x) dx\right)^{\frac{1}{p}} \left[\int_0^\infty \left(\frac{G(y)}{y}\right)^q dy\right]^{\frac{1}{q}},$$

$$\qquad (5.23)$$

$$\left[\int_0^\infty \left(\int_0^\infty \frac{g(y)dy}{x+y}\right)^q dx\right]^{\frac{1}{q}} < \frac{\pi}{p\sin(\pi/p)}$$

$$\times \left[\int_0^\infty \left(\frac{G(y)}{y}\right)^q dy\right]^{\frac{1}{q}}; \qquad (5.24)$$

$$\int_0^\infty \int_0^\infty \frac{f(x)g(y)}{1+xy} dx\, dy$$

$$< \frac{\pi}{p\sin(\pi/p)} \left(\int_0^\infty x^{p-2} f^p(x) dx \right)^{\frac{1}{p}} \left[\int_0^\infty \left(\frac{G(y)}{y} \right)^q dy \right]^{\frac{1}{q}},$$

$$(5.25)$$

$$\left[\int_0^\infty x^{q-2} \left(\int_0^\infty \frac{g(y) dy}{1+xy} \right)^q dx \right]^{\frac{1}{q}} < \frac{\pi}{p\sin(\pi/p)}$$

$$\times \left[\int_0^\infty \left(\frac{G(y)}{y} \right)^q dy \right]^{\frac{1}{q}}.$$

$$(5.26)$$

(ii) for $\lambda = 1, r = p$, and $s = q$, we have the dual forms of (5.23)–(5.26) as follows:

$$\int_0^\infty \int_0^\infty \frac{f(x)g(y)}{x+y} dx\, dy$$

$$< \frac{\pi}{q\sin(\pi/p)} \left(\int_0^\infty x^{p-2} f^p(x) dx \right)^{\frac{1}{p}} \left(\int_0^\infty y^{-2} G^q(y) dy \right)^{\frac{1}{q}},$$

$$(5.27)$$

$$\left[\int_0^\infty x^{q-2} \left(\int_0^\infty \frac{g(y) dy}{x+y} \right)^q dx \right]^{\frac{1}{q}} < \frac{\pi}{q\sin(\pi/p)}$$

$$\times \left(\int_0^\infty y^{-2} G^q(y) dy \right)^{\frac{1}{q}};$$

$$(5.28)$$

$$\int_0^\infty \int_0^\infty \frac{f(x)g(y)}{1+xy} dx\, dy$$

$$< \frac{\pi}{q\sin(\pi/p)} \left(\int_0^\infty f^p(x) dx \right)^{\frac{1}{p}} \left(\int_0^\infty y^{-2} G^q(y) dy \right)^{\frac{1}{q}},$$

$$(5.29)$$

$$\left[\int_0^\infty \left(\int_0^\infty \frac{g(y)dy}{1+xy} \right)^q dx \right]^{\frac{1}{q}} < \frac{\pi}{q\sin(\pi/p)}$$

$$\times \left(\int_0^\infty y^{-2} G^q(y)dy \right)^{\frac{1}{q}}. \tag{5.30}$$

(iii) for $p = q = 2$, both (5.23) and (5.27) reduce to

$$\int_0^\infty \int_0^\infty \frac{f(x)g(y)}{x+y} dx\, dy < \frac{\pi}{2} \left[\int_0^\infty f^2(x)dx \int_0^\infty \left(\frac{G(y)}{y} \right)^2 dy \right]^{\frac{1}{2}} \tag{5.31}$$

and both (5.24) and (5.28) reduce to the following equivalent inequality of (5.31):

$$\left[\int_0^\infty \left(\int_0^\infty \frac{g(y)}{x+y} dy \right)^2 dx \right]^{\frac{1}{2}} < \frac{\pi}{2} \left[\int_0^\infty \left(\frac{G(y)}{y} \right)^2 dy \right]^{\frac{1}{2}}; \tag{5.32}$$

both (5.25) and (5.29) reduce to

$$\int_0^\infty \int_0^\infty \frac{f(x)g(y)}{1+xy} dx\, dy < \frac{\pi}{2} \left[\int_0^\infty f^2(x)dx \int_0^\infty \left(\frac{G(y)}{y} \right)^2 dy \right]^{\frac{1}{2}} \tag{5.33}$$

and both (5.26) and (5.30) reduce to the following equivalent inequality of (5.33):

$$\left[\int_0^\infty \left(\int_0^\infty \frac{g(y)}{1+xy} dy \right)^2 dx \right]^{\frac{1}{2}} < \frac{\pi}{2} \left[\int_0^\infty \left(\frac{G(y)}{y} \right)^2 dy \right]^{\frac{1}{2}}. \tag{5.34}$$

5.3   Operator Expressions

(a) We set

$$\varphi(x) := x^{p(1-\lambda_1)-a-1} \quad \text{and} \quad \psi(y) := y^{-q\lambda_2-a-1},$$

wherefrom

$$\varphi^{1-q}(x) = x^{q(\lambda_1+a)-a-1}(x, y \in \mathbf{R}_+).$$

Define the real normed linear spaces $L_{p,\varphi}(\mathbf{R}_+), L_{q,\psi}(\mathbf{R}_+)$, and $L_{q,\varphi^{1-q}}(\mathbf{R}_+)$ as in Chapter 2.

Assuming that $g(y)$ is a nonnegative measurable function in $\mathbf{R}_+$ such that

$$g \in \widetilde{L}(\mathbf{R}_+) := \{g = g(y); e^{-ty}g(y) \in L(\mathbf{R}_+)\}$$

and setting

$$h(x) := \int_0^\infty \frac{g(y)}{(x+y)^\lambda}dy, x \in \mathbf{R}_+,$$

we can rewrite (5.12) as follows:

$$\|h\|_{q,\varphi^{1-q}} \leq \lambda B^{\frac{1}{p}}(\lambda_2+1,\lambda-\lambda_2)B^{\frac{1}{q}}(\lambda_1,\lambda+1-\lambda_1)\|G\|_{q,\psi} < \infty,$$

namely,

$$h \in L_{q,\varphi^{1-q}}(\mathbf{R}_+).$$

Definition 5.11. Define a Hilbert-type operator

$$T : \widetilde{L}(\mathbf{R}_+) \to L_{q,\varphi^{1-q}}(\mathbf{R}_+)$$

as follows.

For any $g \in \widetilde{L}(\mathbf{R}_+)$, there exists a unique representation $h \in L_{q,\varphi^{1-q}}(\mathbf{R}_+)$ satisfying

$$(Tg)(x) = h(x)$$

for any $x \in \mathbf{R}_+$. Define the formal inner product of $f \in L_{p,\varphi}(\mathbf{R}_+)$ and Tg and the norm of T as follows:

$$(f, Tg) := \int_0^\infty f(x) \left[\int_0^\infty \frac{g(y)}{(x+y)^\lambda} dy \right] dx = I,$$

$$\|T\| = \sup_{g(\neq\theta)\in\tilde{L}(\mathbf{R}_+)} \frac{\|Tg\|_{q,\varphi^{1-q}}}{\|G\|_{q,\psi}}.$$

By Theorem 5.8, we have the following theorem.

Theorem 5.12. *If $f \in L_{p,\varphi}(\mathbf{R}_+), g \in \tilde{L}(\mathbf{R}_+), \|f\|_{p,\varphi}, \|G\|_{q,\psi} > 0$, then we have the following equivalent inequalities:*

$$(f, Tg) < \lambda B^{\frac{1}{p}}(\lambda_2 + 1, \lambda - \lambda_2) B^{\frac{1}{q}}(\lambda_1,$$
$$\times \lambda + 1 - \lambda_1) \|f\|_{p,\varphi} \|G\|_{q,\psi}, \tag{5.35}$$

$$\|Tg\|_{q,\varphi^{1-q}} < \lambda B^{\frac{1}{p}}(\lambda_2 + 1,$$
$$\times \lambda - \lambda_2) B^{\frac{1}{q}}(\lambda_1, \lambda + 1 - \lambda_1) \|G\|_{q,\psi}. \tag{5.36}$$

Moreover, $\lambda_1 + \lambda_2 = \lambda$ if and only if the constant factor

$$\lambda B^{\frac{1}{p}}(\lambda_2 + 1, \lambda - \lambda_2) B^{\frac{1}{q}}(\lambda_1, \lambda + 1 - \lambda_1)$$

in (5.35) and (5.36) is the best possible, namely,

$$\|T\| = \lambda_2 B(\lambda_1, \lambda_2).$$

(b) We set

$$\phi(x) := x^{p(\lambda_1 - \lambda + 1) + a - 1},$$

wherefrom

$$\phi^{1-q}(x) = x^{q(\lambda - \lambda_1 - a) + a - 1} \quad (x \in \mathbf{R}_+).$$

Define the real normed linear spaces $L_{p,\phi}(\mathbf{R}_+)$ and $L_{q,\phi^{1-q}}(\mathbf{R}_+)$ as in Chapter 2.

Assuming that $g(y)$ is a nonnegative measurable function, $g \in \widetilde{L}(\mathbf{R}_+)$, and setting

$$H(x) := \int_0^\infty \frac{g(y)}{(1+xy)^\lambda} dy, x \in \mathbf{R}_+,$$

we can rewrite (5.12) as follows:

$$\|H\|_{q,\phi^{1-q}} \leq \lambda B^{\frac{1}{p}}(\lambda_2+1, \lambda-\lambda_2) B^{\frac{1}{q}}(\lambda_1, \lambda+1-\lambda_1)\|G\|_{q,\psi} < \infty,$$

namely,

$$H \in L_{q,\phi^{1-q}}(\mathbf{R}_+).$$

Definition 5.13. Define a Hilbert-type operator

$$T_1 : \widetilde{L}(\mathbf{R}_+) \to L_{q,\phi^{1-q}}(\mathbf{R}_+)$$

as follows.

For any $g \in \widetilde{L}(\mathbf{R}_+)$, there exists a unique representation $H \in L_{q,\phi^{1-q}}(\mathbf{R}_+)$ satisfying

$$(T_1 g)(x) = H(x)$$

for any $x \in \mathbf{R}_+$. Define the formal inner product of $f \in L_{p,\phi}(\mathbf{R}_+)$ and $T_1 g$ and the norm of T_1 as follows:

$$(f, T_1 g) := \int_0^\infty f(x) \left[\int_0^\infty \frac{g(y)}{(1+xy)^\lambda} dy \right] dx,$$

$$\|T_1\| = \sup_{g(\neq \theta) \in \widetilde{L}(\mathbf{R}_+)} \frac{\|T_1 g\|_{q,\phi^{1-q}}}{\|G\|_{q,\psi}}.$$

By Corollary 5.9, we have the following theorem.

Theorem 5.14. *If* $f \in L_{p,\phi}(\mathbf{R}_+), g \in \widetilde{L}(\mathbf{R}_+), \|f\|_{p,\phi}, \|G\|_{q,\psi} > 0,$ *then we have the following equivalent inequalities:*

$$(f, T_1 g) < \lambda B^{\frac{1}{p}}(\lambda_2+1, \lambda-\lambda_2) B^{\frac{1}{q}}$$
$$\times (\lambda_1, \lambda+1-\lambda_1)\|f\|_{p,\phi}\|G\|_{q,\psi}, \tag{5.37}$$

$$\|T_1 g\|_{q,\phi^{1-q}} < \lambda B^{\frac{1}{p}}(\lambda_2+1, \lambda-\lambda_2) B^{\frac{1}{q}}$$
$$\times (\lambda_1, \lambda+1-\lambda_1)\|G\|_{q,\psi}. \tag{5.38}$$

Moreover, $\lambda_1 + \lambda_2 = \lambda$ if and only if the constant factor

$$\lambda B^{\frac{1}{p}}(\lambda_2 + 1, \lambda - \lambda_2) B^{\frac{1}{q}}(\lambda_1, \lambda + 1 - \lambda_1)$$

in (5.37) and (5.38) is the best possible, namely,

$$\|T_1\| = \lambda_2 B(\lambda_1, \lambda_2).$$

5.4 The Case of Inequality Involving Two Upper-Limit Functions

Hereinafter in this section, we suppose that $p > 1, \frac{1}{p} + \frac{1}{q} = 1, \lambda > 0,$
$\lambda_i \in (0, \lambda)$ $(i = 1, 2)$, $a := \lambda - \lambda_1 - \lambda_2$, $f(x)$ and $g(y)$ are nonnegative
measurable functions in $\mathbf{R}_+$ such that $e^{-tx} f(x)$ and $e^{-ty} g(y)$ $(t > 0)$
are Lebesgue's integrable functions in $\mathbf{R}_+$. We also assume that the
two upper-limit functions

$$F(x) := \int_0^x f(t)dt \quad \text{and} \quad G(y) := \int_0^y g(t)dt \ (x, y \in [0, \infty))$$

satisfy

$$0 < \int_0^\infty x^{-p\lambda_1 - a - 1} F^p(x)dx < \infty \quad \text{and}$$

$$0 < \int_0^\infty y^{-q\lambda_2 - a - 1} G^q(y)dy < \infty.$$

Lemma 5.15 (cf. Lemma 4.1). For $t > 0$, we have the following
equalities:

$$\int_0^\infty e^{-tx} f(x)dx = t \int_0^\infty e^{-tx} F(x)dx, \tag{5.39}$$

$$\int_0^\infty e^{-ty} g(y)dy = t \int_0^\infty e^{-ty} G(y)dy. \tag{5.40}$$

Lemma 5.16. *Define the following weight functions:*

$$\varpi(\lambda_2, x) := x^{\lambda+1-\lambda_2} \int_0^\infty \frac{t^{\lambda_2}}{(x+t)^{\lambda+2}} dt (x \in \mathbf{R}_+), \qquad (5.41)$$

$$\omega(\lambda_1, y) := y^{\lambda+1-\lambda_1} \int_0^\infty \frac{t^{\lambda_1}}{(t+y)^{\lambda+2}} dt (y \in \mathbf{R}_+). \qquad (5.42)$$

We have the following expressions:

$$\varpi(\lambda_2, x) = B(\lambda_2+1, \lambda+1-\lambda_2)(x \in \mathbf{R}_+), \qquad (5.43)$$
$$\omega(\lambda_1, y) = B(\lambda_1+1, \lambda+1-\lambda_1)(y \in \mathbf{R}_+), \qquad (5.44)$$

where

$$B(u, v) = \int_0^\infty \frac{t^{u-1}}{(1+t)^{u+v}} dt \ (u, v > 0)$$

is the beta function.

Proof.   Setting $u = \frac{t}{x}$, we get that

$$\varpi(\lambda_2, x) = x^{\lambda+1-\lambda_2} \int_0^\infty \frac{(ux)^{\lambda_2}}{(x+ux)^{\lambda+2}} x \, du$$

$$= \int_0^\infty \frac{u^{(\lambda_2+1)-1}}{(1+u)^{\lambda+2}} du = B(\lambda_2+1, \lambda+1-\lambda_2),$$

namely, (5.43) follows. Similarly, we have (5.44).
    This completes the proof of the lemma.   □

Lemma 5.17. *We have the following extended Hardy–Hilbert inte-gral inequality:*

$$I_1 := \int_0^\infty \int_0^\infty \frac{F(x)G(y)}{(x+y)^{\lambda+2}} dx \, dy$$

$$< B^{\frac{1}{p}}(\lambda_2+1, \lambda+1-\lambda_2) B^{\frac{1}{q}}(\lambda_1+1, \lambda+1-\lambda_1)$$

$$\times \left(\int_0^\infty x^{-p\lambda_1-a-1} F^p(x) dx \right)^{\frac{1}{p}}$$

$$\times \left(\int_0^\infty y^{-q\lambda_2-a-1} G^q(y) dy \right)^{\frac{1}{q}}. \qquad (5.45)$$

Proof.   By Hölder's inequality (cf. Ref. [120]), we obtain

$$
\begin{aligned}
I_1 &= \int_0^\infty \int_0^\infty \frac{1}{(x+y)^{\lambda+2}} \left(\frac{y^{\lambda_2/p}}{x^{\lambda_1/q}} F(x) \right) \left(\frac{x^{\lambda_1/q}}{y^{\lambda_2/p}} G(y) \right) dx\, dy \\
&\leq \left[\int_0^\infty \int_0^\infty \frac{1}{(x+y)^{\lambda+2}} \frac{y^{\lambda_2}}{x^{\lambda_1(p-1)}} F^p(x) dy\, dx \right]^{\frac{1}{p}} \\
&\quad \times \left[\int_0^\infty \int_0^\infty \frac{1}{(x+y)^{\lambda+2}} \frac{x^{\lambda_1}}{y^{\lambda_2(q-1)}} G^q(y) dx\, dy \right]^{\frac{1}{q}} \\
&= \left(\int_0^\infty \varpi(\lambda_2, x) x^{-p\lambda_1 - a - 1} F^p(x) dx \right)^{\frac{1}{p}} \\
&\quad \times \left(\int_0^\infty \omega(\lambda_1, y) y^{-q\lambda_2 - a - 1} G^q(y) dy \right)^{\frac{1}{q}}.
\end{aligned}
\tag{5.46}
$$

If (5.46) retains the form of equality, then there exist constants A and B such that they are not both zero, satisfying

$$
A \frac{y^{\lambda_2}}{x^{\lambda_1(p-1)}} F^p(x) = B \frac{x^{\lambda_1}}{y^{\lambda_2(q-1)}} G^q(y) \quad \text{a.e. in } \mathbf{R}_+^2.
$$

We assume that $A \neq 0$. Then, there exists a fixed $y \in \mathbf{R}_+$ such that

$$
x^{-p\lambda_1 - a - 1} F^p(x) = \frac{BG^q(y)}{Ay^{q\lambda_2}} x^{-a-1} \quad \text{a.e. in } \mathbf{R}_+.
$$

Since for any $a = \lambda - \lambda_1 - \lambda_2 \in \mathbf{R}$,

$$
\int_0^\infty x^{-a-1} dx = \infty,
$$

the above expression contradicts the fact that

$$
0 < \int_0^\infty x^{-p\lambda_1 - a - 1} F^p(x) dx < \infty.
$$

Therefore, by (5.43) and (5.44), we have (5.45). This completes the proof of the lemma. $\square$

Theorem 5.18. *We have the following Hilbert-type integral inequality involving two upper-limit functions:*

$$I := \int_0^\infty \int_0^\infty \frac{f(x)g(y)}{(x+y)^\lambda} dx\, dy < \lambda(\lambda+1) B^{\frac{1}{p}}(\lambda_2+1, \lambda+1-\lambda_2)$$

$$\times B^{\frac{1}{q}}(\lambda_1+1, \lambda+1-\lambda_1)$$

$$\times \left(\int_0^\infty x^{-p\lambda_1-a-1} F^p(x) dx \right)^{\frac{1}{p}}$$

$$\times \left(\int_0^\infty y^{-q\lambda_2-a-1} G^q(y) dy \right)^{\frac{1}{q}}. \tag{5.47}$$

In particular, for $\lambda_1 + \lambda_2 = \lambda$ (or $a = 0$), we reduce (5.47) to the following:

$$\int_0^\infty \int_0^\infty \frac{f(x)g(y)}{(x+y)^\lambda} dx\, dy < \lambda_1 \lambda_2 B(\lambda_1, \lambda_2)$$

$$\times \left(\int_0^\infty x^{-p\lambda_1-1} F^p(x) dx \right)^{\frac{1}{p}}$$

$$\times \left(\int_0^\infty y^{-q\lambda_2-1} G^q(y) dy \right)^{\frac{1}{q}}, \tag{5.48}$$

where the constant factor $\lambda_1 \lambda_2 B(\lambda_1, \lambda_2)$ is the best possible.

Proof. Using Fubini's theorem (cf. Ref. [119]), (5.39) and (5.40), we deduce that

$$I = \frac{1}{\Gamma(\lambda)} \int_0^\infty \int_0^\infty f(x)g(y) \left[\int_0^\infty e^{-(x+y)t} t^{\lambda-1} dt \right] dx\, dy$$

$$= \frac{1}{\Gamma(\lambda)} \int_0^\infty t^{\lambda-1} \left(\int_0^\infty e^{-xt} f(x) dx \right) \left(\int_0^\infty e^{-ty} g(y) dy \right) dt$$

$$= \frac{1}{\Gamma(\lambda)} \int_0^\infty t^{\lambda-1} \left(t \int_0^\infty e^{-xt} F(x) dx \right) \left(\int_0^\infty t e^{-ty} G(y) dy \right) dt$$

$$= \frac{1}{\Gamma(\lambda)} \int_0^\infty \int_0^\infty F(x)G(y) \left[\int_0^\infty e^{-(x+y)t} t^{(\lambda+2)-1} dt \right] dx\, dy$$

$$= \frac{\Gamma(\lambda+2)}{\Gamma(\lambda)} \int_0^\infty \int_0^\infty \frac{F(x)G(y)}{(x+y)^{\lambda+2}} dx\, dy = \lambda(\lambda+1) I_1. \tag{5.49}$$

Then, by (5.45), we have (5.47).

For $a = 0$ in (5.47), we have (5.48).

For any $0 < \varepsilon < \min\{p\lambda_1, q\lambda_2\}$, we set

$$\widetilde{f}(x) := \begin{cases} 0, & 0 < x \leq 1, \\ x^{\lambda_1 - \frac{\varepsilon}{p} - 1}, & x > 1; \end{cases} \quad \widetilde{g}(y) := \begin{cases} 0, & 0 < y \leq 1, \\ y^{\lambda_2 - \frac{\varepsilon}{q} - 1}, & y > 1. \end{cases}$$

We obtain that

$$e^{-tx}\widetilde{f}(x),\ e^{-ty}\widetilde{g}(y) \in L(\mathbf{R}_+),\ \widetilde{F}(x) = \widetilde{G}(y) = 0 \quad (0 < x, y \leq 1)$$

and

$$\widetilde{F}(x) = \int_1^x t^{\lambda_1 - \frac{\varepsilon}{p} - 1} dt < \int_0^x t^{\lambda_1 - \frac{\varepsilon}{p} - 1} dt$$

$$= \frac{x^{\lambda_1 - \frac{\varepsilon}{p}}}{\lambda_1 - \frac{\varepsilon}{p}} (x > 1),$$

$$\widetilde{G}(y) = \int_1^y t^{\lambda_2 - \frac{\varepsilon}{q} - 1} dt < \int_0^y t^{\lambda_2 - \frac{\varepsilon}{q} - 1} dt$$

$$= \frac{y^{\lambda_2 - \frac{\varepsilon}{q}}}{\lambda_2 - \frac{\varepsilon}{q}} (y > 1).$$

If there exists a positive constant $M(\leq \lambda_1 \lambda_2 B(\lambda_1, \lambda_2))$ such that (5.48) is valid when replacing $\lambda_1 \lambda_2 B(\lambda_1, \lambda_2)$ by M, then in particular, replacing $f(x), g(y), F(x),$ and $G(y)$ by $\widetilde{f}(x), \widetilde{g}(y), \widetilde{F}(x),$ and $\widetilde{G}(y)$, respectively, we have

$$\widetilde{I} := \int_0^\infty \int_0^\infty \frac{\widetilde{f}(x)\widetilde{g}(y)}{(x+y)^\lambda} dx\, dy$$

$$< M \left(\int_0^\infty x^{-p\lambda_1 - 1} \widetilde{F}^p(x) dx \right)^{\frac{1}{p}} \left(\int_0^\infty y^{-q\lambda_2 - 1} \widetilde{G}^q(y) dy \right)^{\frac{1}{q}}$$

$$< \frac{M}{(\lambda_1 - \frac{\varepsilon}{p})(\lambda_2 - \frac{\varepsilon}{q})} \left(\int_1^\infty x^{-\varepsilon - 1} dx \right)^{\frac{1}{p}} \left(\int_1^\infty y^{-\varepsilon - 1} dy \right)^{\frac{1}{q}}$$

$$= \frac{M}{\varepsilon(\lambda_1 - \frac{\varepsilon}{p})(\lambda_2 - \frac{\varepsilon}{q})}.$$

In view of Fubini's theorem (cf. Ref. [119]), setting $u = \frac{y}{x}$, it follows that

$$
\begin{aligned}
\tilde{I} &:= \int_1^\infty x^{\lambda_1 - \frac{\varepsilon}{p} - 1} \left[\int_1^\infty \frac{y^{\lambda_2 - \frac{\varepsilon}{q} - 1}}{(x+y)^\lambda} dy \right] dx \\
&= \int_1^\infty x^{-\varepsilon - 1} \left[\int_{\frac{1}{x}}^\infty \frac{u^{\lambda_2 - \frac{\varepsilon}{q} - 1}}{(1+u)^\lambda} du \right] dx \\
&= \int_1^\infty x^{-\varepsilon - 1} \left[\int_{\frac{1}{x}}^1 \frac{u^{\lambda_2 - \frac{\varepsilon}{q} - 1}}{(1+u)^\lambda} du \right] dx \\
&\quad + \int_1^\infty x^{-\varepsilon - 1} \left[\int_1^\infty \frac{u^{\lambda_2 - \frac{\varepsilon}{q} - 1}}{(1+u)^\lambda} du \right] dx \\
&= \int_0^1 \left(\int_{\frac{1}{u}}^\infty x^{-\varepsilon - 1} dx \right) \frac{u^{\lambda_2 - \frac{\varepsilon}{q} - 1} du}{(1+u)^\lambda} + \frac{1}{\varepsilon} \int_1^\infty \frac{u^{\lambda_2 - \frac{\varepsilon}{q} - 1} du}{(1+u)^\lambda} \\
&= \frac{1}{\varepsilon} \left[\int_0^1 \frac{u^{\lambda_2 + \frac{\varepsilon}{p} - 1}}{(1+u)^\lambda} du + \int_1^\infty \frac{u^{\lambda_2 - \frac{\varepsilon}{q} - 1}}{(1+u)^\lambda} du \right].
\end{aligned}
$$

In virtue of the above results, we obtain

$$
\int_0^1 \frac{u^{\lambda_2 + \frac{\varepsilon}{p} - 1}}{(1+u)^\lambda} du + \int_1^\infty \frac{u^{\lambda_2 - \frac{\varepsilon}{q} - 1}}{(1+u)^\lambda} du
$$

$$
= \varepsilon \tilde{I} < \frac{M}{(\lambda_1 - \frac{\varepsilon}{p})(\lambda_2 - \frac{\varepsilon}{q})}.
$$

For $\varepsilon \to 0^+$ in the above inequality, in view of the continuity of the beta function, we derive that

$$
\frac{M}{\lambda_1 \lambda_2} \geq \int_0^\infty \frac{u^{\lambda_2 - 1}}{(1+u)^\lambda} du = B(\lambda_1, \lambda_2),
$$

namely, $M \geq \lambda_1 \lambda_2 B(\lambda_1, \lambda_2)$. Hence, $M = \lambda_1 \lambda_2 B(\lambda_1, \lambda_2)$ is the best possible constant factor in (5.48).

This completes the proof of the theorem. ☐

Remark 5.19. We set

$$\widehat{\lambda}_1 := \lambda_1 + \frac{a}{p} = \frac{\lambda - \lambda_2}{p} + \frac{\lambda_1}{q} \quad \text{and}$$

$$\widehat{\lambda}_2 := \lambda_2 + \frac{a}{q} = \frac{\lambda - \lambda_1}{q} + \frac{\lambda_2}{p}.$$

It follows that $\widehat{\lambda}_1 + \widehat{\lambda}_2 = \lambda$. We obtain that $0 < \widehat{\lambda}_1 < \frac{\lambda}{p} + \frac{\lambda}{q} = \lambda$ and $0 < \widehat{\lambda}_2 = \lambda - \widehat{\lambda}_1 < \lambda$, and then, $B(\widehat{\lambda}_1 + \widehat{\lambda}_2) \in \mathbf{R}_+$. Thus, we rewrite (5.47) as follows:

$$I = \int_0^\infty \int_0^\infty \frac{f(x)g(y)}{(x+y)^\lambda} dx\, dy$$

$$< \lambda(\lambda+1)B^{\frac{1}{p}}(\lambda_2 + 1, \lambda + 1 - \lambda_2)B^{\frac{1}{q}}(\lambda_1 + 1, \lambda + 1 - \lambda_1)$$

$$\times \left(\int_0^\infty x^{-p\widehat{\lambda}_1 - 1} F^p(x)dx \right)^{\frac{1}{p}} \left(\int_0^\infty y^{-q\widehat{\lambda}_2 - 1} G^q(y)dy \right)^{\frac{1}{q}}.$$

$$(5.50)$$

Theorem 5.20. *If the constant factor*

$$\lambda(\lambda+1)B^{\frac{1}{p}}(\lambda_2 + 1, \lambda + 1 - \lambda_2)B^{\frac{1}{q}}(\lambda_1 + 1, \lambda + 1 - \lambda_1)$$

in (5.47) (or (5.50)) is the best possible, then $a = 0$, *namely,* $\lambda_1 + \lambda_2 = \lambda$.

Proof. By Hölder's inequality (cf. Ref. [120]), we obtain that

$$B(\widehat{\lambda}_1 + 1, \widehat{\lambda}_2 + 1) = \int_0^\infty \frac{u^{\widehat{\lambda}_1}}{(1+u)^{\lambda+2}} du$$

$$= \int_0^\infty \frac{u^{\frac{\lambda - \lambda_2}{p} + \frac{\lambda_1}{q}}}{(1+u)^{\lambda+2}} du$$

$$= \int_0^\infty \frac{1}{(1+u)^{\lambda+2}} \left(u^{\frac{\lambda - \lambda_2}{p}} \right) \left(u^{\frac{\lambda_1}{q}} \right) du$$

$$\leq \left[\int_0^\infty \frac{u^{\lambda - \lambda_2}}{(1+u)^{\lambda+2}} du \right]^{\frac{1}{p}} \left[\int_0^\infty \frac{u^{\lambda_1}}{(1+u)^{\lambda+2}} du \right]^{\frac{1}{q}}$$

$$= B^{\frac{1}{p}}(\lambda_2 + 1, \lambda + 1 - \lambda_2)B^{\frac{1}{q}}(\lambda_1 + 1, \lambda + 1 - \lambda_1). \quad (5.51)$$

By (5.48) (for $\lambda_i = \widehat{\lambda}_i$ ($i = 1, 2$)), since

$$\lambda(\lambda + 1)B^{\frac{1}{p}}(\lambda_2 + 1, \lambda + 1 - \lambda_2)B^{\frac{1}{q}}(\lambda_1 + 1, \lambda + 1 - \lambda_1)$$

is the best possible constant factor in (5.50), we have

$$\lambda(\lambda + 1)B^{\frac{1}{p}}(\lambda_2 + 1, \lambda + 1 - \lambda_2)B^{\frac{1}{q}}(\lambda_1 + 1, \lambda + 1 - \lambda_1)$$
$$\leq \widehat{\lambda}_1\widehat{\lambda}_2 B(\widehat{\lambda}_1, \widehat{\lambda}_2) = \lambda(\lambda + 1)B(\widehat{\lambda}_1 + 1, \widehat{\lambda}_2 + 1)(\in \mathbf{R}_+),$$

namely,

$$B^{\frac{1}{p}}(\lambda_2 + 1, \lambda + 1 - \lambda_2)B^{\frac{1}{q}}(\lambda_1 + 1, \lambda + 1 - \lambda_1)$$
$$\leq B(\widehat{\lambda}_1 + 1, \widehat{\lambda}_2 + 1).$$

It follows that (5.51) retains the form of equality.

We observe that (5.51) retains the form of equality if and only if there exist constants A and B such that they are not both zero and (cf. Ref. [120])

$$Au^{\lambda - \lambda_2} = Bu^{\lambda_1} \quad \text{a.e. in } \mathbf{R}_+.$$

Assuming that $A \neq 0$, we have

$$u^{\lambda - \lambda_1 - \lambda_2} = \frac{B}{A} \quad \text{a.e. in } \mathbf{R}_+.$$

It follows that $a = \lambda - \lambda_1 - \lambda_2 = 0$, namely, $\lambda_1 + \lambda_2 = \lambda$. This completes the proof of the theorem. □

Theorem 5.21. *The following statements, (i), (ii), (iii), and (iv), are equivalent:*

(i) *both* $B^{\frac{1}{p}}(\lambda_2 + 1, \lambda + 1 - \lambda_2)B^{\frac{1}{q}}(\lambda_1 + 1, \lambda + 1 - \lambda_1)$ *and*

$$B\left(\frac{\lambda - \lambda_2}{p} + \frac{\lambda_1}{q} + 1, \frac{\lambda - \lambda_1}{q} + \frac{\lambda_1}{p} + 1\right)$$

are independent of p and q;

(ii) *the following equality holds:*

$$B^{\frac{1}{p}}(\lambda_2 + 1, \lambda + 1 - \lambda_2)B^{\frac{1}{q}}(\lambda_1 + 1, \lambda + 1 - \lambda_1)$$
$$= B\left(\frac{\lambda - \lambda_2}{p} + \frac{\lambda_1}{q} + 1, \frac{\lambda - \lambda_1}{q} + \frac{\lambda_1}{p} + 1\right);$$

(iii) $\lambda_1 + \lambda_2 = \lambda$;

(iv) *the constant factor*

$$\lambda(\lambda + 1)B^{\frac{1}{p}}(\lambda_2 + 1, \lambda + 1 - \lambda_2)B^{\frac{1}{q}}(\lambda_1 + 1, \lambda + 1 - \lambda_1)$$

is the best possible in (5.47).

Proof. (i) $\Rightarrow$ (ii): In view of the continuity of the beta function, we obtain that

$$B^{\frac{1}{p}}(\lambda_2 + 1, \lambda + 1 - \lambda_2)B^{\frac{1}{q}}(\lambda_1 + 1, \lambda + 1 - \lambda_1)$$

$$= \lim_{p \to 1^+} \lim_{q \to \infty} B^{\frac{1}{p}}(\lambda_2 + 1, \lambda + 1 - \lambda_2)B^{\frac{1}{q}}(\lambda_1 + 1, \lambda + 1 - \lambda_1)$$

$$= B(\lambda_2 + 1, \lambda + 1 - \lambda_2),$$

$$B\left(\frac{\lambda - \lambda_2}{p} + \frac{\lambda_1}{q} + 1, \frac{\lambda - \lambda_1}{q} + \frac{\lambda_2}{p} + 1\right)$$

$$= \lim_{p \to 1^+} \lim_{q \to \infty} B\left(\frac{\lambda - \lambda_2}{p} + \frac{\lambda_1}{q} + 1, \frac{\lambda - \lambda_1}{q} + \frac{\lambda_2}{p} + 1\right)$$

$$= B(\lambda_2 + 1, \lambda + 1 - \lambda_2)$$

$$= B^{\frac{1}{p}}(\lambda_2 + 1, \lambda - \lambda_2)B^{\frac{1}{q}}(\lambda_1 + 1, \lambda + 1 - \lambda_1).$$

(ii) $\Rightarrow$ (iii): In this assumption, (5.51) keeps the form of equality. By the proof of Theorem 5.20, we have $\lambda_1 + \lambda_2 = \lambda$.

(iii) $\Rightarrow$ (iv): If $\lambda_1 + \lambda_2 = \lambda$, then by Theorem 5.18, the constant factor

$$\lambda B^{\frac{1}{p}}(\lambda_2 + 1, \lambda - \lambda_2)B^{\frac{1}{q}}(\lambda_1, \lambda + 1 - \lambda_1)(= \lambda B(\lambda_1, \lambda_2))$$

in (5.47) is the best possible.

(iv) $\Rightarrow$ (i): By Theorem 5.20, we have $\lambda_1 + \lambda_2 = \lambda$, and thus,

$$B^{\frac{1}{p}}(\lambda_2 + 1, \lambda + 1 - \lambda_2)B^{\frac{1}{q}}(\lambda_1 + 1, \lambda + 1 - \lambda_1)$$

$$= B\left(\frac{\lambda - \lambda_2}{p} + \frac{\lambda_1}{q} + 1, \frac{\lambda - \lambda_1}{q} + \frac{\lambda_2}{p} + 1\right)$$

$$= B(\lambda_1 + 1, \lambda_2 + 1),$$

both of which are independent of p and q.

Hence, the statements (i), (ii), (iii), and (iv) are equivalent. This completes the proof of the theorem. $\qquad\square$

Remark 5.22. For $a = 0$ in (5.45), we have

$$\int_0^\infty \int_0^\infty \frac{F(x)G(y)}{(x+y)^{\lambda+2}} dx\, dy < B(\lambda_1 + 1, \lambda_2 + 1)$$

$$\times \left(\int_0^\infty x^{-p\lambda_1 - 1} F^p(x) dx \right)^{\frac{1}{p}} \left(\int_0^\infty y^{-q\lambda_2 - 1} G^q(y) dy \right)^{\frac{1}{q}}.$$

$$(5.52)$$

We confirm that the constant factor $B(\lambda_1 + 1, \lambda_2 + 1)$ in (5.52) is the best possible. Otherwise, we would reach a contradiction by (5.49) (for $a = 0$) that the constant factor in (5.48) is not the best possible.

Remark 5.23. For

$$\lambda_1 = \frac{\lambda}{r} \quad \text{and} \quad \lambda_2 = \frac{\lambda}{s} \quad \left(r > 1, \frac{1}{r} + \frac{1}{s} = 1 \right)$$

in (5.48), we have the following integral inequality:

$$\int_0^\infty \int_0^\infty \frac{f(x)g(y)}{(x+y)^\lambda} dx\, dy < \frac{\lambda^2}{rs} B\left(\frac{\lambda}{r}, \frac{\lambda}{s} \right)$$

$$\times \left(\int_0^\infty x^{-\frac{p\lambda}{r} - 1} F^p(x) dx \right)^{\frac{1}{p}} \left(\int_0^\infty y^{-\frac{q\lambda}{s} - 1} G^q(y) dy \right)^{\frac{1}{q}}, \quad (5.53)$$

where the constant factor

$$\frac{\lambda^2}{rs} B\left(\frac{\lambda}{r}, \frac{\lambda}{s} \right)$$

is the best possible.

In particular:

(i) for $\lambda = 1, r = q$, and $s = p$, we have

$$\int_0^\infty \int_0^\infty \frac{f(x)g(y)}{x+y} dx\, dy < \frac{\pi}{pq \sin(\pi/p)} \left[\int_0^\infty \left(\frac{F(x)}{x} \right)^p dx \right]^{\frac{1}{p}}$$

$$\times \left[\int_0^\infty \left(\frac{G(y)}{y} \right)^q dy \right]^{\frac{1}{q}};$$

$$(5.54)$$

(ii) for $\lambda = 1, r = p$, and $s = q$, we have the dual forms of (5.54) as follows:

$$\int_0^\infty \int_0^\infty \frac{f(x)g(y)}{x+y} dx\, dy < \frac{\pi}{pq \sin(\pi/p)} \left(\int_0^\infty x^{-2} F^p(x) dx \right)^{\frac{1}{p}}$$

$$\times \left(\int_0^\infty y^{-2} G^q(y) dy \right)^{\frac{1}{q}} ; \qquad (5.55)$$

(iii) for $p = q = 2$ in (5.54) and (5.55), both of them reduce to

$$\int_0^\infty \int_0^\infty \frac{f(x)g(y)}{x+y} dx\, dy < \frac{\pi}{4}$$

$$\times \left[\int_0^\infty \left(\frac{F(x)}{x} \right)^2 dx \int_0^\infty \left(\frac{G(y)}{y} \right)^2 dy \right]^{\frac{1}{2}} . \qquad (5.56)$$

Chapter 6

Equivalent Properties of Two Kinds of Hardy-Type Integral Inequalities

In this chapter, by introducing independent parameters and using weight functions, we present two kinds of new Hardy-type integral inequalities with a general nonhomogeneous kernel. The equivalent forms related to the best possible constant factor and several parameters are considered. We also deduce the cases of a homogeneous kernel and some particular examples. In the form of applications, the operator expressions, some particular cases involving the extended Riemann zeta function, as well as the reverses are also considered.

6.1 Some Lemmas

Hereinafter in this chapter, we suppose that $h(u)$ is a nonnegative measurable function in $\mathbf{R}_+$, $p > 0$ $(p \neq 1)$, $\frac{1}{p} + \frac{1}{q} = 1$, for $x \in \mathbf{R}_+, I_x^{(1)} := (0, x), I_x^{(2)} := (x, \infty), \gamma = \sigma, \sigma_1 \in \mathbf{R} = (-\infty, \infty)$ such that

$$k_j(\gamma) := \int_{I_1^{(j)}} h(u)u^{\gamma-1}du \in \mathbf{R}_+ (j = 1, 2). \tag{6.1}$$

We assume that $f(x)$ and $g(y)$ are nonnegative measurable functions in $\mathbf{R}_+$, satisfying

$$0 < \int_0^\infty x^{p[1-(\frac{\sigma}{p}+\frac{\sigma_1}{q})]-1} f^p(x) dx < \infty \quad \text{and}$$

$$0 < \int_0^\infty y^{q[1-(\frac{\sigma}{p}+\frac{\sigma_1}{q})]-1} g^q(y) dy < \infty,$$

and denote

$$F_j(y) := \int_{I_{y-1}^{(j)}} h(xy)f(x)dx, \quad G_j(x) := \int_{I_{x-1}^{(j)}} h(xy)g(y)dy \quad (j = 1, 2),$$

where

$$I_{y-1}^{(1)} = \left(0, \frac{1}{y}\right) \quad \text{and} \quad I_{y-1}^{(2)} = \left(\frac{1}{y}, \infty\right).$$

Lemma 6.1 (cf. Lemma 2.1). If there exists a constant $\delta_0 > 0$ such that $k_j(\sigma \pm \delta_0) < \infty$ $(j = 1, 2)$, then for $j = 1, 2$, the function $k_j(\eta)$ is continuous in any $\eta \in (\sigma - \delta_0, \sigma + \delta_0)$.

Lemma 6.2. *For $p > 1$ $(q > 1), j = 1, 2$, we have the following Hardy-type integral inequality with a nonhomogeneous kernel:*

$$H_j := \int_0^\infty G_j(x)f(x)dx = \int_0^\infty F_j(y)g(y)dy$$

$$< k_j^{\frac{1}{p}}(\sigma)(\sigma)k_j^{\frac{1}{q}}(\sigma_1) \left\{ \int_0^\infty x^{p[1-(\frac{\sigma}{p}+\frac{\sigma_1}{q})]-1} f^p(x) dx \right\}^{\frac{1}{p}}$$

$$\times \left\{ \int_0^\infty y^{q[1-(\frac{\sigma}{p}+\frac{\sigma_1}{q})]-1} g^q(y) dy \right\}^{\frac{1}{q}}. \tag{6.2}$$

Proof. For $\gamma = \sigma, \sigma_1$, $j = 1, 2$, we define the following weight function:

$$w^{(j)}(\gamma, x) := x^\gamma \int_{I_{x-1}^{(j)}} h(xy)y^{\gamma-1}dy(x \in \mathbf{R}_+). \tag{6.3}$$

Setting $u = xy$, we obtain

$$w^{(j)}(\gamma, x) = \int_{I_1^{(j)}} h(u)u^{\gamma-1}du = k_j(\gamma). \tag{6.4}$$

By Hölder's inequality and Fubini's theorem (cf. Refs. [119,120]), in view of (6.3), we have

$$
H_j = \int_0^\infty \int_{I_{x-1}^{(j)}} h(xy) \left[\frac{y^{(\sigma-1)/p}}{x^{(\sigma_1-1)/q}} f(x) \right] \left[\frac{x^{(\sigma_1-1)/q}}{y^{(\sigma-1)/p}} g(y) \right] dy\, dx
$$

$$
\leq \left\{ \int_0^\infty \int_{I_{x-1}^{(j)}} h(xy) \frac{y^{\sigma-1}}{x^{(\sigma_1-1)(p-1)}} f^p(x) dy\, dx \right\}^{\frac{1}{p}}
$$

$$
\times \left\{ \int_0^\infty \int_{I_{x-1}^{(j)}} h(xy) \frac{x^{\sigma_1-1}}{y^{(\sigma-1)(q-1)}} g^q(y) dy\, dx \right\}^{\frac{1}{q}}
$$

$$
= \left\{ \int_0^\infty \left[\int_{I_{x-1}^{(j)}} h(xy) \frac{y^{\sigma-1}}{x^{(\sigma_1-1)(p-1)}} dy \right] f^p(x) dx \right\}^{\frac{1}{p}}
$$

$$
\times \left\{ \int_0^\infty \left[\int_{I_{y-1}^{(j)}} h(xy) \frac{x^{\sigma_1-1}}{y^{(\sigma-1)(q-1)}} dx \right] g^q(y) dy \right\}^{\frac{1}{q}}
$$

$$
= \left\{ \int_0^\infty \omega^{(j)}(\sigma, x) x^{p[1-(\frac{\sigma}{p}+\frac{\sigma_1}{q})]-1} f^p(x) dx \right\}^{\frac{1}{p}}
$$

$$
\times \left\{ \int_0^\infty \omega^{(j)}(\sigma_1, y) y^{q[1-(\frac{\sigma}{p}+\frac{\sigma_1}{q})]-1} g^q(y) dy \right\}^{\frac{1}{q}}. \tag{6.5}
$$

Setting

$$
E^{(j)} := \left\{ (x, y); x > 0, y \in I_{x-1}^{(j)} \right\} \quad (j = 1, 2),
$$

if (6.5) retains the form of equality, then there exist constants A and B such that they are not both zero and (cf. Ref. [120])

$$
A \frac{y^{\sigma-1}}{x^{(\sigma_1-1)(p-1)}} f^p(x) = B \frac{x^{\sigma_1-1}}{y^{(\sigma-1)(q-1)}} g^q(y) \quad \text{a.e. in } E^{(j)}.
$$

Assuming that $A \neq 0$, there exists a fixed $y \in I_{x-1}^{(j)}$ such that

$$
x^{p[1-(\frac{\sigma}{p}+\frac{\sigma_1}{q})]-1} f^p(x) = \frac{B g^q(y)}{A y^{(\sigma-1)q}} x^{\sigma_1-\sigma-1} \quad \text{a.e. in } (0, \infty).
$$

In view of the fact that

$$\int_0^\infty x^{\sigma_1-\sigma-1}dx = \infty,$$

the above expression contradicts the fact that

$$0 < \int_0^\infty x^{p[1-(\frac{\sigma}{p}+\frac{\sigma_1}{q})]-1}f^p(x)dx < \infty.$$

By (6.4) and (6.5), we have (6.2).

This completes the proof of the lemma. □

Remark 6.3. (i) If $\sigma_1 = \sigma$, then by (6.2), for $p > 1, j = 1,2$, we have

$$0 < \int_0^\infty x^{p(1-\sigma)-1}f^p(x)dx < \infty, \quad 0 < \int_0^\infty y^{q(1-\sigma)-1}g^q(y)dy < \infty,$$

and the following inequality:

$$H_j := \int_0^\infty G_j(x)f(x)dx = \int_0^\infty F_j(y)g(y)dy$$

$$< k_j(\sigma)\left[\int_0^\infty x^{p(1-\sigma)-1}f^p(x)dx\right]^{\frac{1}{p}}\left[\int_0^\infty y^{q(1-\sigma)-1}g^q(y)dy\right]^{\frac{1}{q}}.$$

$$(6.6)$$

(ii) For $0 < p < 1$ $(q < 0), j = 1,2$, by the reverse Hölder inequality (cf. Ref. [120]), similarly, we have the reverses of (6.2) and (6.6).

Lemma 6.4. For $p > 1, j = 1,2$, the constant factor $k_j(\sigma)$ in (6.6) is the best possible.

Proof. We only prove the case of $j = 2$. For any $0 < \varepsilon < p\delta_0$, we set

$$\widetilde{f}(x) := \begin{cases} 0, & 0 < x < 1, \\ x^{\sigma-\frac{\varepsilon}{p}-1}, & x \geq 1; \end{cases} \quad \widetilde{g}(y) := \begin{cases} y^{\sigma+\frac{\varepsilon}{q}-1}, & 0 < y \leq 1, \\ 0, & y > 1. \end{cases}$$

If there exists a positive constant $M(\leq k_2(\sigma))$ such that (6.6) is valid when replacing $k_2(\sigma)$ by M, then in particular, we have

$$\tilde{H}_2 := \int_0^\infty \tilde{g}(y) \left(\int_{\frac{1}{y}}^\infty h(xy)\tilde{f}(x)dx \right) dy$$

$$< M \left[\int_0^\infty x^{p(1-\sigma)-1} \tilde{f}^p(x)dx \right]^{\frac{1}{p}} \left[\int_0^\infty y^{q(1-\sigma)-1} \tilde{g}^q(y)dy \right]^{\frac{1}{q}}$$

$$= M \left(\int_1^\infty x^{-\varepsilon-1}dx \right)^{\frac{1}{p}} \left(\int_0^1 y^{\varepsilon-1}dy \right)^{\frac{1}{q}} = \frac{M}{\varepsilon}.$$

We also obtain that

$$\tilde{H}_2 = \int_0^1 \left(\int_{\frac{1}{y}}^\infty h(xy)x^{\sigma-\frac{\varepsilon}{p}-1}dx \right) y^{\sigma+\frac{\varepsilon}{q}-1}dy$$

$$\overset{u=xy}{=} \int_0^1 \left(\int_1^\infty h(u)u^{\sigma-\frac{\varepsilon}{p}-1}du \right) y^{\varepsilon-1}dy = \frac{1}{\varepsilon}k_2\left(\sigma - \frac{\varepsilon}{p}\right). \quad (6.7)$$

In view of the above results, it follows that

$$k_2\left(\sigma - \frac{\varepsilon}{p}\right) = \varepsilon\tilde{H}_2 < M.$$

For $\varepsilon \to 0^+$, by Fatou's lemma (cf. Ref. [119]), we obtain that

$$k_2(\sigma) = \int_1^\infty \lim_{\varepsilon\to0^+} h(u)u^{\sigma-\frac{\varepsilon}{p}-1}du \leq \lim_{\varepsilon\to0^+} \int_1^\infty h(u)u^{\sigma-\frac{\varepsilon}{p}-1}du \leq M.$$

Hence, $M = k_2(\sigma)$ is the best possible constant factor in (6.6) (for $j = 2$).

Similarly, setting $\tilde{f}(x)$ and $\tilde{g}(y)$ as follows:

$$\tilde{f}(x) := \begin{cases} y^{\sigma+\frac{\varepsilon}{p}-1}, & 0 < x \leq 1, \\ 0, & x > 1; \end{cases} \quad \tilde{g}(y) := \begin{cases} 0, & 0 < y < 1, \\ y^{\sigma-\frac{\varepsilon}{q}-1}, & y \geq 1, \end{cases}$$

we can prove the case for $j = 1$.

This completes the proof of the lemma. $\qquad\square$

Remark 6.5. We set

$$\widehat{\sigma} := \frac{\sigma}{p} + \frac{\sigma_1}{q}.$$

We can rewrite (6.2) as follows.
For $j = 1, 2$,

$$\int_0^\infty G_j(x) f(x) dx = \int_0^\infty F_j(y) g(y) dy$$

$$< k_j^{\frac{1}{p}}(\sigma) k_j^{\frac{1}{q}}(\sigma_1) \left[\int_0^\infty x^{p(1-\widehat{\sigma})-1} f^p(x) dx \right]^{\frac{1}{p}}$$

$$\times \left[\int_0^\infty y^{q(1-\widehat{\sigma})-1} g^q(y) dy \right]^{\frac{1}{q}}. \tag{6.8}$$

By Hölder's inequality with weight (cf. Ref. [120]), we have the following inequality:

$$0 < k_j(\widehat{\sigma}) = k_j \left(\frac{\sigma}{p} + \frac{\sigma_1}{q} \right) = \int_{I_1^{(j)}} h(u) u^{\frac{\sigma}{p} + \frac{\sigma_1}{q} - 1} du$$

$$= \int_{I_1^{(j)}} h(u) (u^{\frac{\sigma-1}{p}})(u^{\frac{\sigma_1-1}{q}}) du$$

$$\leq \left(\int_{I_1^{(j)}} h(u) u^{\sigma-1} du \right)^{\frac{1}{p}} \left(\int_{I_1^{(j)}} h(u) u^{\sigma_1-1} du \right)^{\frac{1}{q}}$$

$$= k_j^{\frac{1}{p}}(\sigma) k_j^{\frac{1}{q}}(\sigma_1) < \infty \quad (j = 1, 2). \tag{6.9}$$

Lemma 6.6. *For $p > 1, j = 1, 2$, if the constant factor*

$$k_j^{\frac{1}{p}}(\sigma) k_j^{\frac{1}{q}}(\sigma_1)$$

in (6.8) (or (6.2)) is the best possible, then $\sigma_1 = \sigma$.

Proof. If the constant factor $k_j^{\frac{1}{p}}(\sigma) k_j^{\frac{1}{q}}(\sigma_1)$ in (6.8) is the best possible, then in view of (6.6) (for $\sigma = \widehat{\sigma}$), we have

$$k_j^{\frac{1}{p}}(\sigma) k_j^{\frac{1}{q}}(\sigma_1) \leq k_j(\widehat{\sigma}) \ (\in \mathbf{R}_+),$$

namely, (6.9) retains the form of equality.

We observe that (6.9) retains the form of equality if and only if that there exist constants A and B such that they are not both zero and (cf. Ref. [120])

$$Au^{\sigma-1} = Bu^{\sigma_1-1} \quad \text{a.e. in } I_1^{(j)}.$$

Assuming that $A \neq 0$, we have

$$u^{\sigma-\sigma_1} = \frac{B}{A} \quad \text{a.e. in } I_1^{(j)},$$

which follows that $\sigma - \sigma_1 = 0$, namely, $\sigma_1 = \sigma$.
The lemma is proved. □

6.2 Main Results and Some Corollaries

Theorem 6.7. *For $p > 1, j = 1, 2$, inequality (6.2) is equivalent to the following inequality:*

$$K_j := \left[\int_0^\infty y^{p(\frac{\sigma}{p}+\frac{\sigma_1}{q})-1} F_j^p(y) dy \right]^{\frac{1}{p}}$$

$$< k_j^{\frac{1}{p}}(\sigma) k_j^{\frac{1}{q}}(\sigma_1) \left\{ \int_0^\infty x^{p[1-(\frac{\sigma}{p}+\frac{\sigma_1}{q})]-1} f^p(x) dx \right\}^{\frac{1}{p}}. \quad (6.10)$$

The constant factor

$$k_j^{\frac{1}{p}}(\sigma) k_j^{\frac{1}{q}}(\sigma_1)$$

in (6.10) is the best possible if and only if the same constant factor in (6.2) is the best possible.

In particular, for $\sigma_1 = \sigma, j = 1, 2$, we have the following inequality equivalent to (6.6) with the same best possible constant factor $k_j(\sigma)$:

$$\left(\int_0^\infty y^{p\sigma-1} F_j^p(y) dy \right)^{\frac{1}{p}} < k_j(\sigma) \left[\int_0^\infty x^{p(1-\sigma)-1} f^p(x) dx \right]^{\frac{1}{p}}. \quad (6.11)$$

Proof. If (6.10) is valid, then by Hölder's inequality, we have

$$H_j = \int_0^\infty \left[y^{\frac{-1}{p}+(\frac{\sigma}{p}+\frac{\sigma_1}{q})} F_j(y) \right] \left[y^{\frac{1}{p}-(\frac{\sigma}{p}+\frac{\sigma_1}{q})} g(y) \right] dy$$

$$\leq K_j \left\{ \int_0^\infty y^{q[1-(\frac{\sigma}{p}+\frac{\sigma_1}{q})]-1} g^q(y) dy \right\}^{\frac{1}{q}}. \quad (6.12)$$

By (6.10), we deduce (6.2).

On the other hand, assuming that (6.2) is valid, we set

$$g(y) := y^{p(\frac{\sigma}{p}+\frac{\sigma_1}{q})-1} F_j^{p-1}(y)(y > 0).$$

Then, it follows that

$$K_j^p = \int_0^\infty y^{q[1-(\frac{\sigma}{p}+\frac{\sigma_1}{q})]-1} g^q(y) dy = H_j. \tag{6.13}$$

If $K_j = 0$, then (6.10) is naturally valid; if $K_j = \infty$, then it is impossible to make (6.10) valid, namely, $K_j < \infty$. Suppose that $0 < K_j < \infty$. By (6.2), we have

$$0 < K_j^p = \int_0^\infty y^{q[1-(\frac{\sigma}{p}+\frac{\sigma_1}{q})]-1} g^q(y) dy = H_j$$

$$< k_j^{\frac{1}{p}}(\sigma) k_j^{\frac{1}{q}}(\sigma_1) \left\{ \int_0^\infty x^{p[1-(\frac{\sigma}{p}+\frac{\sigma_1}{q})]-1} f^p(x) dx \right\}^{\frac{1}{p}} K_j^{p-1} < \infty,$$

$$K_j = \left\{ \int_0^\infty y^{q[1-(\frac{\sigma}{p}+\frac{\sigma_1}{q})]-1} g^q(y) dy \right\}^{\frac{1}{p}}$$

$$< k_j^{\frac{1}{p}}(\sigma) k_j^{\frac{1}{q}}(\sigma_1) \left\{ \int_0^\infty x^{p[1-(\frac{\sigma}{p}+\frac{\sigma_1}{q})]-1} f^p(x) dx \right\}^{\frac{1}{p}},$$

namely, (6.10) follows, which is equivalent to (6.2).

If the constant factor in (6.2) is the best possible, then the same constant factor in (6.10) is also the best possible. Otherwise, by (6.12), we would reach a contradiction that the same constant factor in (6.2) is not the best possible. Similarly, if the constant factor in (6.10) is the best possible, then the same constant factor in (6.2) is also the best possible. Otherwise, by (6.13), we would reach a contradiction that the same constant factor in (6.10) is not the best possible.

This completes the proof of the theorem.   □

Theorem 6.8. *For $p > 1, j = 1, 2$, the following statements,* (i), (ii), (iii), *and* (iv), *are equivalent:*

(i) *both* $k_j^{\frac{1}{p}}(\sigma) k_j^{\frac{1}{q}}(\sigma_1)$ *and* $k_j(\frac{\sigma}{p} + \frac{\sigma_1}{q})$ *are independent of p, qj;*

(ii) $k_j^{\frac{1}{p}}(\sigma) k_j^{\frac{1}{q}}(\sigma_1) \leq k_j(\frac{\sigma}{p} + \frac{\sigma_1}{q})$;

(iii) $\sigma_1 = \sigma$;

(iv) the constant factor $k_j^{\frac{1}{p}}(\sigma)k_j^{\frac{1}{q}}(\sigma_1)$ in (6.2) and (6.10) is the best possible.

Proof. (i) $\Rightarrow$ (ii): We have

$$k_j^{\frac{1}{p}}(\sigma)k_j^{\frac{1}{q}}(\sigma_1) = \lim_{p\to 1^+}\lim_{q\to\infty} k_j^{\frac{1}{p}}(\sigma)k_j^{\frac{1}{q}}(\sigma_1) = k_j(\sigma).$$

By Fatou's lemma (cf. Ref. [119]), we derive that

$$k_j\left(\frac{\sigma}{p} + \frac{\sigma_1}{q}\right) = \lim_{p\to 1^+}\lim_{q\to\infty} k_j\left(\frac{\sigma}{p} + \frac{\sigma_1}{q}\right)$$

$$\geq k_j(\sigma) = k_j^{\frac{1}{p}}(\sigma)k_j^{\frac{1}{q}}(\sigma_1).$$

(ii) $\Rightarrow$ (iii): If

$$k_j^{\frac{1}{p}}(\sigma)k_j^{\frac{1}{q}}(\sigma_1) \leq k_j\left(\frac{\sigma}{p} + \frac{\sigma_1}{q}\right),$$

then (6.9) retains the form of equality. In view of the proof of Lemma 6.6, we have $\sigma_1 = \sigma$.

(iii) $\Rightarrow$ (iv): By Lemma 6.4 and Theorem 6.7, the constant factor $k_j^{\frac{1}{p}}(\sigma)k_j^{\frac{1}{q}}(\sigma_1)(= k_j(\sigma))$ is the best possible in (6.2) and (6.10).

(iv) $\Rightarrow$ (i): By Lemma 6.6, we have $\sigma_1 = \sigma$, and then, both

$$k_j^{\frac{1}{p}}(\sigma)k_j^{\frac{1}{q}}(\sigma_1) \quad \text{and} \quad k_j\left(\frac{\sigma}{p} + \frac{\sigma_1}{q}\right)$$

are equal to $k_j(\sigma)$, which are independent of p and q.

Hence, statements (i), (ii), (iii), and (iv) are equivalent.

This completes the proof of the theorem. $\qquad\square$

If $k_\lambda(x, y)(\geq 0)$ is a homogeneous function of degree $-\lambda$, satisfying

$$k_\lambda(ux, uy) = u^{-\lambda}k_\lambda(x, y) \ (u, x, y > 0),$$

then setting $h(u) = k_\lambda(1, u)$, replacing x by $\frac{1}{x}$ and $x^{\lambda-2}f(\frac{1}{x})$ by $f(x)$ in Lemma 6.2 and Theorems 6.7 and 6.8, for $\sigma_1 = \lambda - \mu$,

$$k_{\lambda,j}(\gamma) := \int_{I_1^{(j)}} k_\lambda(1, u)u^{\sigma-1}du \in \mathbf{R}_+$$

$(\gamma = \sigma, \lambda - \mu)$, and setting

$$\widetilde{F}_j(y) := \int_{I_y^{(j)}} k_\lambda(x, y) f(x) dx,$$

$$\widetilde{G}_j(x) := \int_{I_x^{(j)}} k_\lambda(x, y) g(y) dy (j = 1, 2),$$

we deduce the following corollary.

Corollary 6.9. *For $p > 1, j = 1, 2,$*

$$0 < \int_0^\infty x^{p[1-(\frac{\lambda-\sigma}{p}+\frac{\mu}{q})]-1} f^p(x) dx < \infty \quad and$$

$$0 < \int_0^\infty y^{q[1-(\frac{\sigma}{p}+\frac{\lambda-\mu}{q})]-1} g^q(y) dy < \infty,$$

then we have the following equivalent Hardy-type integral inequalities with a homogeneous kernel:

$$\int_0^\infty \widetilde{G}_j(x) f(x) dx = \int_0^\infty \widetilde{F}_j(y) g(y) dy$$

$$< k_{\lambda,j}^{\frac{1}{p}}(\sigma) k_{\lambda,j}^{\frac{1}{q}}(\lambda - \mu) \left\{ \int_0^\infty x^{p[1-(\frac{\lambda-\sigma}{p}+\frac{\mu}{q})]-1} f^p(x) dx \right\}^{\frac{1}{p}}$$

$$\times \left\{ \int_0^\infty y^{q[1-(\frac{\sigma}{p}+\frac{\lambda-\mu}{q})]-1} g^q(y) dy \right\}^{\frac{1}{q}}, \tag{6.14}$$

$$\left[\int_0^\infty y^{p(\frac{\sigma}{p}+\frac{\lambda-\mu}{q})-1} \widetilde{F}_j^p(y) dy \right]^{\frac{1}{p}}$$

$$< k_{\lambda,j}^{\frac{1}{p}}(\sigma) k_{\lambda,j}^{\frac{1}{q}}(\lambda - \mu) \left\{ \int_0^\infty x^{p[1-(\frac{\lambda-\sigma}{p}+\frac{\mu}{q})]-1} f^p(x) dx \right\}^{\frac{1}{p}}. \tag{6.15}$$

Moreover, the constant factor

$$k_{\lambda,j}^{\frac{1}{p}}(\sigma) k_{\lambda,j}^{\frac{1}{q}}(\lambda - \mu)$$

is the best possible in (6.14) *if and only if the same constant factor in* (6.15) *is the best possible.*

In particular, for $\mu + \sigma = \lambda$, $j = 1, 2$, we have the following equivalent inequalities with the best possible constant factor $k_{\lambda,j}(\sigma)$:

$$\int_0^\infty \widetilde{G}_j(x)f(x)dx = \int_0^\infty \widetilde{F}_j(y)g(y)dy$$

$$< k_{\lambda,j}(\sigma)\left[\int_0^\infty x^{p(1-\mu)-1}f^p(x)dx\right]^{\frac{1}{p}}\left[\int_0^\infty y^{q(1-\sigma)-1}g^q(y)dy\right]^{\frac{1}{q}},$$

$$\tag{6.16}$$

$$\left(\int_0^\infty y^{p\sigma-1}\widetilde{F}_j^p(y)dy\right)^{\frac{1}{p}} < k_{\lambda,j}(\sigma)\left[\int_0^\infty x^{p(1-\mu)-1}f^p(x)dx\right]^{\frac{1}{p}}.$$

$$\tag{6.17}$$

Corollary 6.10. *For $p > 1, j = 1, 2$, the following statements, (I), (II), (III), and (IV), are equivalent:*

(I) *both*

$$k_{\lambda,j}^{\frac{1}{p}}(\sigma)k_{\lambda,j}^{\frac{1}{q}}(\lambda - \mu) \quad \text{and} \quad k_{\lambda,j}\left(\frac{\sigma}{p} + \frac{\lambda - \mu}{q}\right)$$

are independent of p and q;

(II) $k_{\lambda,j}^{\frac{1}{p}}(\sigma)k_\lambda^{\frac{1}{q}}(\lambda - \mu) \le k_{\lambda,j}(\frac{\sigma}{p} + \frac{\lambda-\mu}{q})$;

(III) $\mu + \sigma = \lambda$;

(IV) *the constant factor $k_{\lambda,j}^{\frac{1}{p}}(\sigma)k_{\lambda,j}^{\frac{1}{q}}(\lambda - \mu)$ in (6.14) and (6.15) is the best possible.*

Example 6.11. (i) For $\sigma > 0, j = 1, h(u) = 1$,

$$k_1(\sigma) = \int_0^1 u^{\sigma-1}du = \frac{1}{\sigma},$$

in (6.6) and (6.15), we have the following equivalent Hardy-type integral inequalities with the best possible constant factor $\frac{1}{\sigma}$:

$$\int_0^\infty \left(\int_0^y f(x)dx\right) g(y)dy$$

$$< \frac{1}{\sigma}\left[\int_0^\infty x^{p(1-\sigma)-1}f^p(x)dx\right]^{\frac{1}{p}}\left[\int_0^\infty y^{q(1-\sigma)-1}g^q(y)dy\right]^{\frac{1}{q}}, \quad (6.18)$$

$$\left[\int_0^\infty y^{p\sigma-1}\left(\int_0^{\frac{1}{y}} f(x)dx\right)^p dy\right]^{\frac{1}{p}} < \frac{1}{\sigma}\left[\int_0^\infty x^{p(1-\sigma)-1}f^p(x)dx\right]^{\frac{1}{p}}.$$

(6.19)

(ii) For $\sigma < 0, j = 2, h(u) = 1$,

$$k_2(\sigma) = \int_1^\infty u^{\sigma-1}du = \frac{-1}{\sigma},$$

in (6.6) and (6.15), replacing σ by $-\sigma$, we have the following equivalent Hardy-type integral inequalities with the best possible constant factor $\frac{1}{\sigma}$ $(\sigma > 0)$:

$$\int_0^\infty \left(\int_{\frac{1}{y}}^\infty f(x)dx\right) g(y)dy$$

$$< \frac{1}{\sigma}\left[\int_0^\infty x^{p(1+\sigma)-1}f^p(x)dx\right]^{\frac{1}{p}}\left[\int_0^\infty y^{q(1+\sigma)-1}g^q(y)dy\right]^{\frac{1}{q}}, \quad (6.20)$$

$$\left[\int_0^\infty y^{-p\sigma-1}\left(\int_{\frac{1}{y}}^\infty f(x)dx\right)^p dy\right]^{\frac{1}{p}} < \frac{1}{\sigma}\left[\int_0^\infty x^{p(1+\sigma)-1}f^p(x)dx\right]^{\frac{1}{p}}.$$

(6.21)

(iii) For $\sigma > 0, j = 1, k_\lambda(1, u) = 1$,

$$k_{\lambda,1}(\sigma) = \int_0^1 u^{\sigma-1}du = \frac{1}{\sigma},$$

$$k_\lambda(x, y) = x^{-\lambda}k_\lambda\left(1, \frac{y}{x}\right) = x^{-\lambda}(\lambda \in \mathbf{R}),$$

in (6.16) and (6.17), we have the following equivalent Hardy-type integral inequalities with the best possible constant factor $\frac{1}{\sigma}(\mu = \lambda - \sigma)$:

$$\int_0^\infty \left(\int_0^y x^{-\lambda}f(x)dx\right) g(y)dy$$

$$< \frac{1}{\sigma}\left[\int_0^\infty x^{p(1+\sigma-\lambda)-1}f^p(x)dx\right]^{\frac{1}{p}}\left[\int_0^\infty y^{q(1-\sigma)-1}g^q(y)dy\right]^{\frac{1}{q}},$$

(6.22)

$$\left[\int_0^\infty y^{p\sigma-1} \left(\int_0^y x^{-\lambda} f(x) dx \right)^p dy \right]^{\frac{1}{p}}$$

$$< \frac{1}{\sigma} \left[\int_0^\infty x^{p(1+\sigma-\lambda)-1} f^p(x) dx \right]^{\frac{1}{p}}. \tag{6.23}$$

(iv) For $\sigma < 0, j = 2, k_\lambda(1, u) = 1$,

$$k_{\lambda,2}(\sigma) = \int_1^\infty u^{\sigma-1} du = \frac{-1}{\sigma},$$

$$k_\lambda(x, y) = x^{-\lambda} k_\lambda \left(1, \frac{y}{x} \right) = x^{-\lambda} \ (\lambda \in \mathbf{R}),$$

in (6.16) and (6.17), replacing σ by $-\sigma$ ($\mu = \lambda + \sigma$), we have the following equivalent Hardy-type integral inequalities with the best possible constant factor $\frac{1}{\sigma}$ ($\sigma > 0$):

$$\int_0^\infty \left(\int_y^\infty x^{-\lambda} f(x) dx \right) g(y) dy$$

$$< \frac{1}{\sigma} \left[\int_0^\infty x^{p(1-\sigma-\lambda)-1} f^p(x) dx \right]^{\frac{1}{p}} \left[\int_0^\infty y^{q(1+\sigma)-1} g^q(y) dy \right]^{\frac{1}{q}}, \tag{6.24}$$

$$\left[\int_0^\infty y^{-p\sigma-1} \left(\int_y^\infty x^{-\lambda} f(x) dx \right)^p dy \right]^{\frac{1}{p}}$$

$$< \frac{1}{\sigma} \left[\int_0^\infty x^{p(1-\sigma-\lambda)-1} f^p(x) dx \right]^{\frac{1}{p}}. \tag{6.25}$$

6.3   Operator Expressions and Some Particular Cases

(a) For $p > 1$, we set

$$\varphi(x) := x^{p[1-(\frac{\sigma}{p}+\frac{\sigma_1}{q})]-1} \quad \text{and} \quad \psi(y) := y^{q[1-(\frac{\sigma}{p}+\frac{\sigma_1}{q})]-1},$$

wherefrom

$$\psi^{1-p}(y) = y^{p(\frac{\sigma}{p}+\frac{\sigma_1}{q})-1}(x, y \in \mathbf{R}_+).$$

As in Chapter 2, we define the real normed linear spaces $L_{p,\varphi}(\mathbf{R}_+)$, $L_{q,\psi}(\mathbf{R}_+)$, and $L_{p,\psi^{1-p}}(\mathbf{R}_+)$.

For $f \in L_{p,\varphi}(\mathbf{R}_+), j = 1, 2$, we may rewrite (6.10) as follows:

$$\|F_j\|_{p,\psi^{1-p}} = \left[\int_0^\infty \psi^{1-p}(y) F_j^p(y) dy \right]^{\frac{1}{p}}$$

$$< k_j^{\frac{1}{p}}(\sigma) k_j^{\frac{1}{q}}(\sigma_1) \|f\|_{p,\varphi} < \infty, \qquad (6.26)$$

namely, $F_j \in L_{p,\psi^{1-p}}(\mathbf{R}_+)$.

Definition 6.12. Define a Hardy-type integral operator with the nonhomogeneous kernel

$$T_j \ : \ L_{p,\varphi}(\mathbf{R}_+) \to L_{p,\psi^{1-p}}(\mathbf{R}_+)$$

as follows.

For any $f \in L_{p,\varphi}(\mathbf{R}_+)$, there exists a unique representation

$$F_j = T_j f \in L_{p,\psi^{1-p}}(\mathbf{R}_+)$$

satisfying

$$T_j f(y) = F_j(y)$$

for any $y \in \mathbf{R}_+$.

Define the formal inner product of $T_j f$ and $g \in L_{q,\psi}(\mathbf{R}_+)$ and the norm of T_j as follows:

$$(T_j f, g) := \int_0^\infty F_j(y) g(y) dy = H_j,$$

$$\|T_j\| = \sup_{f(\neq \theta) \in L_{p,\varphi}(\mathbf{R}_+)} \frac{\|T_j f\|_{p,\psi^{1-p}}}{\|f\|_{p,\varphi}}.$$

In view of (6.26), it follows that

$$\|T_j f\|_{p,\psi^{1-p}} = \|F_j\|_{p,\psi^{1-p}} \leq k_j^{\frac{1}{p}}(\sigma) k_j^{\frac{1}{q}}(\sigma_1) \|f\|_{p,\varphi},$$

and then, the operator T_j is bounded satisfying

$$||T_j|| \leq k_j^{\frac{1}{p}}(\sigma)k_j^{\frac{1}{q}}(\sigma_1).$$

By Theorems 6.7 and 6.8, we have the following theorem.

Theorem 6.13. *If $p > 1$, $f(>0) \in L_{p,\varphi}(\mathbf{R}_+), g(>0) \in L_{q,\psi}(\mathbf{R}_+)$, then for $j = 1, 2$, we have the following equivalent inequalities:*

$$(T_j f, g) < k_j^{\frac{1}{p}}(\sigma)k_j^{\frac{1}{q}}(\sigma_1)||f||_{p,\varphi}||g||_{q,\psi}, \tag{6.27}$$

$$||T_j f||_{p,\psi^{1-p}} < k_j^{\frac{1}{p}}(\sigma)k_j^{\frac{1}{q}}(\sigma_1)||f||_{p,\varphi}. \tag{6.28}$$

Moreover, $\sigma_1 = \sigma$ if and only if the constant factor

$$k_j^{\frac{1}{p}}(\sigma)k_j^{\frac{1}{q}}(\sigma_1)$$

in (6.27) and (6.28) is the best possible, namely, $||T_j|| = k_j(\sigma)$.

(b) For $p > 1$, we set

$$\Phi(x) := x^{p[1-(\frac{\lambda-\sigma}{p}+\frac{\mu}{q})]-1} \quad \text{and} \quad \Psi(y) := y^{q[1-(\frac{\sigma}{p}+\frac{\lambda-\mu}{q})]-1},$$

wherefrom

$$\Psi^{1-p}(y) = y^{p(\frac{\sigma}{p}+\frac{\lambda-\mu}{q})-1}(x, y \in \mathbf{R}_+).$$

As in Chapter 2, we define the real normed linear spaces $L_{p,\Phi}(\mathbf{R}_+)$, $L_{q,\Psi}(\mathbf{R}_+)$, and $L_{p,\Psi^{1-p}}(\mathbf{R}_+)$.

For $f \in L_{p,\Phi}(\mathbf{R}_+), j = 1, 2$, we may rewrite (6.15) as follows:

$$||\widetilde{F}_j||_{p,\Psi^{1-p}} = \left(\int_0^\infty \Psi^{1-p}(y)\widetilde{F}_j^p(y)dy \right)^{\frac{1}{p}}$$

$$< k_{\lambda,j}^{\frac{1}{p}}(\sigma)k_{\lambda,j}^{\frac{1}{q}}(\lambda - \mu)||f||_{p,\Phi} < \infty, \tag{6.29}$$

namely, $\widetilde{F}_j \in L_{p,\Psi^{1-p}}(\mathbf{R}_+)$.

Definition 6.14. Define a Hardy-type integral operator with the homogeneous kernel

$$\widetilde{T}_j \; : \; L_{p,\Phi}(\mathbf{R}_+) \to L_{p,\Psi^{1-p}}(\mathbf{R}_+)$$

as follows.

For any $f \in L_{p,\Phi}(\mathbf{R}_+)$, there exists a unique representation

$$\widetilde{F}_j = \widetilde{T}_j f \in L_{p,\Psi^{1-p}}(\mathbf{R}_+)$$

satisfying

$$\widetilde{T}_j f(y) = \widetilde{F}_j(y)$$

for any $y \in \mathbf{R}_+$.

Define the formal inner product of $\widetilde{T}_j f$ and $g \in L_{q,\Psi}(\mathbf{R}_+)$ and the norm of $\widetilde{T}_j$ as follows:

$$(\widetilde{T}_j f, g) := \int_0^\infty \widetilde{F}_j(y) g(y) dy,$$

$$\|\widetilde{T}_j\| = \sup_{f(\neq\theta)\in L_{p,\Phi}(\mathbf{R}_+)} \frac{\|\widetilde{T}_j f\|_{p,\Psi^{1-p}}}{\|f\|_{p,\Phi}}.$$

In view of (6.29), it follows that

$$\|\widetilde{T}_j f\|_{p,\Psi^{1-p}} = \|\widetilde{F}_j\|_{p,\Psi^{1-p}} \leq k_{\lambda,j}^{\frac{1}{p}}(\sigma) k_{\lambda,j}^{\frac{1}{q}}(\lambda - \mu) \|f\|_{p,\Phi},$$

and then, the operator $\widetilde{T}_j$ is bounded satisfying

$$\|\widetilde{T}_j\| \leq k_{\lambda,j}^{\frac{1}{p}}(\sigma) k_{\lambda,j}^{\frac{1}{q}}(\lambda - \mu).$$

By Corollaries 6.9 and 6.10, we deduce the following corollary.

Corollary 6.15. *For $p > 1, j = 1, 2$, if $f(>0) \in L_{p,\Phi}(\mathbf{R}_+)$, $g(>0) \in L_{q,\Psi}(\mathbf{R}_+)$, then we have the following equivalent inequalities:*

$$(\widetilde{T}_j f, g) < k_{\lambda,j}^{\frac{1}{p}}(\sigma) k_{\lambda,j}^{\frac{1}{q}}(\lambda - \mu) \|f\|_{p,\Phi} \|g\|_{q,\Psi}, \tag{6.30}$$

$$\|\widetilde{T}_j f\|_{p,\Psi^{1-p}} < k_{\lambda,j}^{\frac{1}{p}}(\sigma) k_{\lambda,j}^{\frac{1}{q}}(\lambda - \mu) \|f\|_{p,\Phi}. \tag{6.31}$$

Moreover, $\mu + \sigma = \lambda$ if and only if the constant factor

$$k_{\lambda,j}^{\frac{1}{p}}(\sigma)k_{\lambda,j}^{\frac{1}{q}}(\lambda - \mu)$$

in (6.30) and (6.31) is the best possible, namely,

$$\|\widetilde{T}_j\| = k_{\lambda,j}(\sigma).$$

Example 6.16. (i) Setting

$$h(u) = k_\lambda(1, u) = \frac{(\min\{1, u\})^\gamma |\ln u|^\beta}{|u^{\lambda+\gamma} - 1|} \quad (\beta > 0, \lambda > -\gamma),$$

we get that

$$h(xy) = \frac{(\min\{1, xy\})^\gamma |\ln(xy)|^\beta}{|(xy)^{\lambda+\gamma} - 1|},$$

$$k_\lambda(x, y) = \frac{(\min\{x, y\})^\gamma |\ln(x/y)|^\beta}{|x^{\lambda+\gamma} - y^{\lambda+\gamma}|} \quad (x, y > 0).$$

For $\sigma, \mu = \lambda - \sigma \in (-\gamma, \lambda)$, we obtain

$$k_1(\sigma) = k_{\lambda,1}(\sigma) = \int_0^1 \frac{(\min\{1, u\})^\gamma |\ln u|^\beta}{|u^{\lambda+\gamma} - 1|} u^{\sigma-1} du$$

$$= \int_0^1 \frac{u^\gamma(-\ln u)^\beta}{1 - u^{\lambda+\gamma}} u^{\sigma-1} du = \int_0^1 \frac{(-\ln u)^\beta}{1 - u^{\lambda+\gamma}} u^{\sigma+\gamma-1} du$$

$$= \int_0^1 (-\ln u)^\beta \sum_{k=0}^\infty u^{k(\lambda+\gamma)} u^{\sigma+\gamma-1} du.$$

By Lebesgue's term-by-term integration theorem (cf. Ref. [119]), we have

$$k_1(\sigma) = k_{\lambda,1}(\sigma) = \sum_{k=0}^\infty \int_0^1 (-\ln u)^\beta u^{k(\lambda+\gamma)+\sigma+\gamma-1} du$$

$$= \int_0^\infty v^\beta e^{-v} dv \sum_{k=0}^\infty \frac{1}{[k(\lambda + \gamma) + \sigma + \gamma]^{\beta+1}}$$

$$= \frac{\Gamma(\beta + 1)}{(\lambda + \gamma)^{\beta+1}} \zeta(\beta + 1, \frac{\sigma + \gamma}{\lambda + \gamma}),$$

where

$$\zeta(s,a) := \sum_{k=0}^{\infty} \frac{1}{(k+a)^s} \quad (\mathrm{Re}\, s > 1, 0 < a \leq 1)$$

is the Hurwitz zeta function (with $\zeta(s,1) = \zeta(s)$ being the Riemann zeta function) (cf. Ref. [123]).

Then, by Theorem 6.13 and Corollary 6.15, we have

$$||T_1|| = ||\tilde{T}_1|| = \frac{\Gamma(\beta+1)}{(\lambda+\gamma)^{\beta+1}}\zeta\left(\beta+1, \frac{\sigma+\gamma}{\lambda+\gamma}\right).$$

Similarly, we have

$$k_2(\sigma) = k_{\lambda,2}(\sigma) = \frac{\Gamma(\beta+1)}{(\lambda+\gamma)^{\beta+1}}\zeta\left(\beta+1, \frac{\mu+\gamma}{\lambda+\gamma}\right),$$

and then,

$$||T_2|| = ||\tilde{T}_2|| = \frac{\Gamma(\beta+1)}{(\lambda+\gamma)^{\beta+1}}\zeta\left(\beta+1, \frac{\mu+\gamma}{\lambda+\gamma}\right).$$

(ii) Setting $h(u) = k_\lambda(1,u) = \frac{(\max\{1,u\})^\gamma |\ln u|^\beta}{|u^{\lambda+\gamma}-1|}$ $(\beta > 0, \lambda > -\gamma)$, we find

$$h(xy) = \frac{(\max\{1, xy\})^\gamma |\ln(xy)|^\beta}{|(xy)^{\lambda+\gamma} - 1|},$$

$$k_\lambda(x,y) = \frac{(\max\{x, y\})^\gamma |\ln(x/y)|^\beta}{|x^{\lambda+\gamma} - y^{\lambda+\gamma}|} \quad (x, y > 0).$$

For $\sigma, \mu = \lambda - \sigma \in (0, \lambda + \gamma)$, we obtain

$$k_1(\sigma) = k_{\lambda,1}(\sigma) = \int_0^\infty \frac{(\max\{1, u\})^\gamma |\ln u|^\beta}{|u^{\lambda+\gamma} - 1|} u^{\sigma-1} du$$

$$= \int_0^1 \frac{(-\ln u)^\beta}{1 - u^{\lambda+\gamma}} u^{\sigma-1} du = \int_0^1 \frac{(-\ln u)^\beta}{1 - u^{\lambda+\gamma}} u^{\sigma-1} du$$

$$= \int_0^1 (-\ln u)^\beta \sum_{k=0}^\infty u^{k(\lambda+\gamma)} u^{\sigma-1} du.$$

By Lebesgue's term-by-term integration theorem (cf. Ref. [119]), we have

$$k_1(\sigma) = k_{\lambda,1}(\sigma) = \sum_{k=0}^{\infty} \int_0^1 (-\ln u)^\beta u^{k(\lambda+\gamma)+\sigma-1} du$$

$$= \int_0^\infty v^\beta e^{-v} dv \sum_{k=0}^{\infty} \frac{1}{[k(\lambda+\gamma)+\sigma]^{\beta+1}}$$

$$= \frac{\Gamma(\beta+1)}{(\lambda+\gamma)^{\beta+1}} \zeta\left(\beta+1, \frac{\sigma}{\lambda+\gamma}\right).$$

Then, by Theorem 6.13 and Corollary 6.15, we deduce that

$$\|T_1\| = \|\widetilde{T}_1\| = \frac{\Gamma(\beta+1)}{(\lambda+\gamma)^{\beta+1}} \zeta\left(\beta+1, \frac{\sigma}{\lambda+\gamma}\right).$$

Similarly, we have

$$k_2(\sigma) = k_{\lambda,2}(\sigma) = \frac{\Gamma(\beta+1)}{(\lambda+\gamma)^{\beta+1}} \zeta\left(\beta+1, \frac{\mu}{\lambda+\gamma}\right),$$

and then,

$$\|T_2\| = \|\widetilde{T}_2\| = \frac{\Gamma(\beta+1)}{(\lambda+\gamma)^{\beta+1}} \zeta\left(\beta+1, \frac{\mu}{\lambda+\gamma}\right).$$

6.4 The Case of Reverses

Lemma 6.17. *For $0 < p < 1$ $(q < 0), j = 1, 2$, if there exists a constant $\delta_0 > 0$ such that $k_j(\sigma \pm \delta_0) < \infty$, then the constant factor in the reverse of (6.6),*

$$H_j := \int_0^\infty G_j(x) f(x) dx = \int_0^\infty F_j(y) g(y) dy$$

$$> k_j(\sigma) \left[\int_0^\infty x^{p(1-\sigma)-1} f^p(x) dx\right]^{\frac{1}{p}} \left[\int_0^\infty y^{q(1-\sigma)-1} g^q(y) dy\right]^{\frac{1}{q}},$$

$$\tag{6.32}$$

is the best possible.

Proof.   In the case of $j = 2$, for any $0 < \varepsilon < p\delta_0$, we set

$$\widetilde{f}(x) := \begin{cases} 0, & 0 < x < 1, \\ x^{\sigma - \frac{\varepsilon}{p} - 1}, & x \geq 1; \end{cases} \qquad \widetilde{g}(y) := \begin{cases} y^{\sigma + \frac{\varepsilon}{q} - 1}, & 0 < y \leq 1, \\ 0, & y > 1. \end{cases}$$

If there exists a positive constant M ($\geq k_2(\sigma)$) such that (6.32) is valid when replacing $k_2(\sigma)$ by M, then in particular, we have

$$\widetilde{H}_2 := \int_0^\infty \left(\int_{\frac{1}{y}}^\infty h(xy)\widetilde{f}(x)dx \right) \widetilde{g}(y)dy$$

$$> M \left[\int_0^\infty x^{p(1-\sigma)-1} \widetilde{f}^p(x)dx \right]^{\frac{1}{p}} \left[\int_0^\infty y^{q(1-\sigma)-1} \widetilde{g}^q(y)dy \right]^{\frac{1}{q}}$$

$$= M \left(\int_1^\infty x^{-\varepsilon-1}dx \right)^{\frac{1}{p}} \left(\int_0^1 y^{\varepsilon-1}dy \right)^{\frac{1}{q}} = \frac{M}{\varepsilon}.$$

By (6.4) (for $j = 2$), in view of

$$\sigma - \frac{\varepsilon}{p} \in (\sigma - \delta_0, \sigma + \delta_0),$$

we derive that

$$\widetilde{H}_j = \int_0^1 \left(\int_{\frac{1}{y}}^\infty h(xy)x^{\sigma - \frac{\varepsilon}{p} - 1}dx \right) y^{\sigma + \frac{\varepsilon}{q} - 1}dy$$

$$= \int_0^1 \omega^{(2)} \left(\sigma - \frac{\varepsilon}{p}, y \right) y^{\varepsilon-1}dy = k_2 \left(\sigma - \frac{\varepsilon}{p} \right) \int_0^1 y^{\varepsilon-1}dy$$

$$= \frac{1}{\varepsilon} k_2 \left(\sigma - \frac{\varepsilon}{p} \right) < \infty.$$

In view of the above results, it follows that

$$\infty > k_2 \left(\sigma - \frac{\varepsilon}{p} \right) \geq \varepsilon \widetilde{I} > M.$$

For $\varepsilon \to 0^+$, by Lemma 6.1, we get that $k_2(\sigma) \geq M$. Hence, $M = k_2(\sigma)$ is the best possible constant factor in (6.32) (for $j = 2$). We can similarly treat the case of $j = 1$.

This completes the proof of the lemma.   □

Remark 6.18. For $0 < p < 1, j = 1, 2$, we set $\widehat{\sigma} := \frac{\sigma}{p} + \frac{\sigma_1}{q} = \sigma - \frac{\sigma - \sigma_1}{q}$. We can rewrite the reverse of (6.2) as follows:

$$H_j := \int_0^\infty G_j(x)f(x)dx = \int_0^\infty F_j(y)g(y)dy$$

$$> k_j^{\frac{1}{p}}(\sigma)k_j^{\frac{1}{q}}(\sigma_1)\left[\int_0^\infty x^{p(1-\widehat{\sigma})-1}f^p(x)dx\right]^{\frac{1}{p}}$$

$$\times \left[\int_0^\infty y^{q(1-\widehat{\sigma})-1}g^q(y)dy\right]^{\frac{1}{q}}. \tag{6.33}$$

By the reverse Hölder inequality with weight (cf. Ref. [120]), we have the following reverse inequality:

$$k_j(\widehat{\sigma}) = \int_{I_1^{(j)}} h(u)(u^{\frac{\sigma-1}{p}})(u^{\frac{\sigma_1-1}{q}})du$$

$$\geq \left(\int_{I_1^{(j)}} h(u)u^{\sigma-1}du\right)^{\frac{1}{p}}\left(\int_{I_1^{(j)}} h(u)u^{\sigma_1-1}du\right)^{\frac{1}{q}}$$

$$= k_j^{\frac{1}{p}}(\sigma)k_j^{\frac{1}{q}}(\sigma_1) > 0. \tag{6.34}$$

If $\sigma - \sigma_1 \in (q\delta_0, -q\delta_0)$, then $\widehat{\sigma} \in (\sigma - \delta_0, \sigma + \delta_0)$, and then,

$$0 < k_j(\widehat{\sigma}) = k_j\left(\sigma - \frac{\sigma - \sigma_1}{q}\right) < \infty.$$

Lemma 6.19. *For $0 < p < 1, j = 1, 2$ if there exists a constant $\delta_0 > 0$ such that $k_j(\sigma \pm \delta_0) < \infty$, the constant factor*

$$k_j^{\frac{1}{p}}(\sigma)k_j^{\frac{1}{q}}(\sigma_1)$$

in (6.33) (or the reverse of (6.2)) is the best possible. Then, for $\sigma - \sigma_1 \in (q\delta_0, -q\delta_0)$, we have $\sigma_1 = \sigma$.

Proof. If the constant factor $k_j^{\frac{1}{p}}(\sigma)k_j^{\frac{1}{q}}(\sigma_1)$ in (6.33) is the best possible, then in view of the reverse of (6.6) (for $\sigma = \widehat{\sigma}$), we have

$$k_j^{\frac{1}{p}}(\sigma)k_j^{\frac{1}{q}}(\sigma_1) \geq k_j(\widehat{\sigma}) \ (\in \mathbf{R}_+),$$

namely, (6.34) retains the form of equality.

We observe that (6.34) retains the form of equality if and only if there exist constants A and B such that they are not both zero and (cf. Ref. [120])

$$Au^{\sigma-1} = Bu^{\sigma_1-1} \quad \text{a.e. in } I_1^{(j)}.$$

Assuming that $A \neq 0$, we have

$$u^{\sigma-\sigma_1} = \frac{B}{A} \quad \text{a.e. in } I_1^{(j)},$$

which yields that $\sigma - \sigma_1 = 0$, namely, $\sigma_1 = \sigma$.

This completes the proof of the lemma.   □

Theorem 6.20. *For $0 < p < 1, j = 1, 2$, the reverse of (6.2) is equivalent to the following inequalities:*

$$K_j := \left[\int_0^\infty y^{p(\frac{\sigma}{p}+\frac{\sigma_1}{q})-1} F_j^p(y) dy \right]^{\frac{1}{p}}$$

$$> k_j^{\frac{1}{p}}(\sigma) k_j^{\frac{1}{q}}(\sigma_1) \left\{ \int_0^\infty x^{p[1-(\frac{\sigma}{p}+\frac{\sigma_1}{q})]-1} f^p(x) dx \right\}^{\frac{1}{p}}, \quad (6.35)$$

$$L_j := \left[\int_0^\infty x^{q(\frac{\sigma}{p}+\frac{\sigma_1}{q})-1} G_j^q(x) dx \right]^{\frac{1}{q}}$$

$$> k_j^{\frac{1}{p}}(\sigma) k_j^{\frac{1}{q}}(\sigma_1) \left\{ \int_0^\infty y^{q[1-(\frac{\sigma}{p}+\frac{\sigma_1}{q})]-1} g^q(y) dy \right\}^{\frac{1}{q}}. \quad (6.36)$$

The constant factor

$$k_j^{\frac{1}{p}}(\sigma) k_j^{\frac{1}{q}}(\sigma_1)$$

in (6.35) and (6.36) is the best possible if and only if the same constant factor in the reverse of (6.2) is the best possible.

In particular, for $\sigma_1 = \sigma, j = 1, 2$, if there exists a constant $\delta_0 > 0$ such that $k_j(\sigma \pm \delta_0) < \infty$, then we have the following reverse Hardy-type inequalities equivalent to (6.32) with the same best possible constant factor $k_j(\sigma)$:

$$\left(\int_0^\infty y^{p\sigma-1} F_j^p(y) dy\right)^{\frac{1}{p}} > k_j(\sigma) \left[\int_0^\infty x^{p(1-\sigma)-1} f^p(x) dx\right]^{\frac{1}{p}}, \quad (6.37)$$

$$\left(\int_0^\infty x^{q\sigma-1} G_j^q(x) dx\right)^{\frac{1}{q}} > k_j(\sigma) \left[\int_0^\infty y^{q(1-\sigma)-1} g^q(y) dy\right]^{\frac{1}{q}}. \quad (6.38)$$

Proof. If (6.35) is valid, then by the reverse Hölder inequality (cf. Ref. [120]), we have

$$H_j = \int_0^\infty \left[y^{\frac{-1}{p}+(\frac{\sigma}{p}+\frac{\sigma_1}{q})} F_j(y)\right] \left[y^{\frac{1}{p}-(\frac{\sigma}{p}+\frac{\sigma_1}{q})} g(y)\right] dy$$

$$\geq K_j \left\{\int_0^\infty y^{q[1-(\frac{\sigma}{p}+\frac{\sigma_1}{q})]-1} g^q(y) dy\right\}^{\frac{1}{q}}. \quad (6.39)$$

By (6.35), we deduce the reverse of (6.2).

On the other hand, assuming that the reverse of (6.2) is valid, we set

$$g(y) := y^{p(\frac{\sigma}{p}+\frac{\sigma_1}{q})-1} F_j^{p-1}(y)(y > 0).$$

Then, it follows that

$$K_j^p = \int_0^\infty y^{q[1-(\frac{\sigma}{p}+\frac{\sigma_1}{q})]-1} g^q(y) dy = H_j. \quad (6.40)$$

If $K_j = \infty$, then (6.35) is naturally valid; if $K_j = 0$, then it is impossible to make (6.35) valid, namely, $K_j > 0$. Suppose that $0 < K_j < \infty$. By the reverse of (6.2), we have

$$\infty > K_j^p = \int_0^\infty y^{q[1-(\frac{\sigma}{p}+\frac{\sigma_1}{q})]-1} g^q(y) dy = H_j$$

$$> k_j^{\frac{1}{p}}(\sigma) k_j^{\frac{1}{q}}(\sigma_1) \left\{\int_0^\infty x^{p[1-(\frac{\sigma}{p}+\frac{\sigma_1}{q})]-1} f^p(x) dx\right\}^{\frac{1}{p}} K_j^{p-1} > 0,$$

$$K_j = \left\{ \int_0^\infty y^{q[1-(\frac{\sigma}{p}+\frac{\sigma_1}{q})]-1} g^q(y) dy \right\}^{\frac{1}{p}}$$

$$> k_j^{\frac{1}{p}}(\sigma) k_j^{\frac{1}{q}}(\sigma_1) \left\{ \int_0^\infty x^{p[1-(\frac{\sigma}{p}+\frac{\sigma_1}{q})]-1} f^p(x) dx \right\}^{\frac{1}{p}},$$

namely, (6.35) follows, which is equivalent to the reverse of (6.2).

Similarly, we can show that (6.36) is equivalent to the reverse of (6.2). Hence, the reverse of (6.2), the inequalities (6.35) and (6.36) are equivalent.

If the constant factor $k_j^{\frac{1}{p}}(\sigma) k_j^{\frac{1}{q}}(\sigma_1)$ in the reverse of (6.2) is the best possible, then the same constant factor in (6.35) is also the best possible. Otherwise, by (6.39), we would reach a contradiction that the same constant factor in the reverse (6.2) is not the best possible.

If the constant factor $k_j^{\frac{1}{p}}(\sigma) k_j^{\frac{1}{q}}(\sigma_1)$ in (6.35) is the best possible, then the same constant factor in the reverse of (6.2) is also the best possible. Otherwise, by (6.40), we would reach a contradiction that the same constant factor in (6.35) is not the best possible.

In the same way, we can show that the constant factor $k_j^{\frac{1}{p}}(\sigma) k_j^{\frac{1}{q}}(\sigma_1)$ in the reverse of (6.2) is the best possible if and only if the same constant factor in (6.36) is the best possible. Therefore, the constant factor $k_j^{\frac{1}{p}}(\sigma) k_j^{\frac{1}{q}}(\sigma_1)$ in (6.35) and (6.36) is the best possible if and only if the same constant factor in the reverse of (6.2) is the best possible.

This completes the proof of the theorem. $\square$

Theorem 6.21. *For $0 < p < 1, j = 1, 2$, if there exists a constant $\delta_0 > 0$ such that $k_j(\sigma \pm \delta_0) < \infty$, then the following statements, (i), (ii), (iii), and (iv), are equivalent:*

(i) *both*

$$k_j^{\frac{1}{p}}(\sigma) k_j^{\frac{1}{q}}(\sigma_1) \quad \text{and} \quad k_j \left(\frac{\sigma}{p} + \frac{\sigma_1}{q} \right)$$

are independent of p and q;

(ii) $k_j^{\frac{1}{p}}(\sigma) k_j^{\frac{1}{q}}(\sigma_1) = k_j(\frac{\sigma}{p} + \frac{\sigma_1}{q})$;

(iii) *for $\sigma - \sigma_1 \in (q\delta_0, -q\delta_0)$, we have $\sigma_1 = \sigma$;*

(iv) the constant factor $k_j^{\frac{1}{p}}(\sigma)k_j^{\frac{1}{q}}(\sigma_1)$ in the reverse of (6.2), inequalities (6.35) and (6.36), is the best possible.

Proof. (i) $\Rightarrow$ (ii): We have

$$k_j^{\frac{1}{p}}(\sigma)k_j^{\frac{1}{q}}(\sigma_1) = \lim_{p\to 1^-}\lim_{q\to-\infty} k_j^{\frac{1}{p}}(\sigma)k_j^{\frac{1}{q}}(\sigma_1) = k_j(\sigma).$$

By Lemma 6.1, we deduce that

$$k_j\left(\frac{\sigma}{p}+\frac{\sigma_1}{q}\right) = \lim_{p\to 1^-}\lim_{q\to-\infty} k_j\left(\sigma+\frac{\sigma_1-\sigma}{q}\right)$$

$$= k_j(\sigma) = k_j^{\frac{1}{p}}(\sigma)k_j^{\frac{1}{q}}(\sigma_1).$$

(ii) $\Rightarrow$ (iii): If

$$k_j^{\frac{1}{p}}(\sigma)k_j^{\frac{1}{q}}(\sigma_1) = k_j\left(\frac{\sigma}{p}+\frac{\sigma_1}{q}\right),$$

then (6.34) retains the form of equality. In view of the proof of Lemma 6.19, for $\sigma-\sigma_1 \in (q\delta_0, -q\delta_0)$, we have $\sigma_1 = \sigma$.

(iii) $\Rightarrow$ (i): For $\sigma_1 = \sigma$, both

$$k_j^{\frac{1}{p}}(\sigma)k_j^{\frac{1}{q}}(\sigma_1) \quad \text{and} \quad k_j\left(\frac{\sigma}{p}+\frac{\sigma_1}{q}\right)$$

are independent of p and q, which are equal to $k_j(\sigma)$.

Hence, we have (i) $\Leftrightarrow$ (ii) $\Leftrightarrow$ (iii).

(iii) $\Rightarrow$ (iv): For $\sigma_1 = \sigma$, by Theorem 6.20, the constant factor

$$k_j^{\frac{1}{p}}(\sigma)k_j^{\frac{1}{q}}(\sigma_1)(= k_j(\sigma))$$

is the best possible in the reverse of (6.2), inequalities (6.35) and (6.36).

(iv) $\Rightarrow$ (iii): Since

$$\sigma-\sigma_1 \in (q\delta_0, -q\delta_0),$$

by Lemma 6.19, we have $\sigma_1 = \sigma$.

Hence, we have $(iii) \Leftrightarrow (iv)$.

Therefore, the statements (i), (ii), (iii), and (iv) are equivalent. This completes the proof of the theorem. $\qquad\square$

If $k_\lambda(x,y)(\geq 0)$ is a homogeneous function of degree $-\lambda$ satisfying

$$k_\lambda(ux, uy) = u^{-\lambda}k_\lambda(x,y)(u, x, y > 0),$$

then setting $h(u) = k_\lambda(1, u)$, replacing x by $\frac{1}{x}$ and $x^{\lambda-2}f(\frac{1}{x})$ by $f(x)$ in Theorems 6.19 and 6.20, for $\sigma_1 = \lambda - \mu$,

$$k_{\lambda,j}(\gamma) := \int_{I_1^{(j)}} k_\lambda(1, u)u^{\sigma-1}du \in \mathbf{R}_+ \ (\gamma = \sigma, \lambda - \mu),$$

and setting

$$\widetilde{F}_j(y) = \int_{I_y^{(j)}} k_\lambda(x, y)f(x)dx,$$

$$\widetilde{G}_j(x) = \int_{I_x^{(j)}} k_\lambda(x, y)g(y)dy(j = 1, 2),$$

we have the following theorem.

Corollary 6.22. *For $0 < p < 1, j = 1, 2$,*

$$0 < \int_0^\infty x^{p[1-(\frac{\lambda-\sigma}{p}+\frac{\mu}{q})]-1}f^p(x)dx < \infty, \quad and$$

$$0 < \int_0^\infty y^{q[1-(\frac{\sigma}{p}+\frac{\lambda-\mu}{q})]-1}g^q(y)dy < \infty,$$

we have the following equivalent reverse Hardy-type integral inequalities with a homogeneous kernel:

$$\int_0^\infty \widetilde{G}_j(x)f(x)dx = \int_0^\infty \widetilde{F}_j(y)g(y)dy$$

$$> k_{\lambda,j}^{\frac{1}{p}}(\sigma)k_{\lambda,j}^{\frac{1}{q}}(\lambda - \mu)\left\{\int_0^\infty x^{p[1-(\frac{\lambda-\sigma}{p}+\frac{\mu}{q})]-1}f^p(x)dx\right\}^{\frac{1}{p}}$$

$$\times \left\{\int_0^\infty y^{q[1-(\frac{\sigma}{p}+\frac{\lambda-\mu}{q})]-1}g^q(y)dy\right\}^{\frac{1}{q}}, \tag{6.41}$$

$$\left[\int_0^\infty y^{p(\frac{\sigma}{p}+\frac{\lambda-\mu}{q})-1}\widetilde{F}_j^p(y)dy\right]^{\frac{1}{p}}$$

$$> k_{\lambda,j}^{\frac{1}{p}}(\sigma)k_{\lambda,j}^{\frac{1}{q}}(\lambda - \mu)\left\{\int_0^\infty x^{p[1-(\frac{\lambda-\sigma}{p}+\frac{\mu}{q})]-1}f^p(x)dx\right\}^{\frac{1}{p}}, \tag{6.42}$$

$$\left[\int_0^\infty x^{q(\frac{\lambda-\sigma}{p}+\frac{\mu}{q})-1} \widetilde{G}_j^q(x)dx \right]^{\frac{1}{q}}$$

$$> k_{\lambda,j}^{\frac{1}{p}}(\sigma) k_{\lambda,j}^{\frac{1}{q}}(\lambda-\mu) \left\{ \int_0^\infty y^{q[1-(\frac{\sigma}{p}+\frac{\lambda-\mu}{q})]-1} g^q(y)dy \right\}^{\frac{1}{q}}. \tag{6.43}$$

Moreover, the constant factor

$$k_{\lambda,j}^{\frac{1}{p}}(\sigma) k_{\lambda,j}^{\frac{1}{q}}(\lambda-\mu)$$

is the best possible in (6.41) *if and only if the same constant factor in* (6.42) *and* (6.43) *is the best possible.*

In particular, for $\mu + \sigma = \lambda$, $j = 1,2$, if there exists a constant $\delta_0 > 0$ such that $k_{\lambda,j}(\sigma \pm \delta_0) < \infty$, then we have the following equivalent inequalities with the best possible constant factor $k_{\lambda,j}(\sigma)$:

$$\int_0^\infty \widetilde{G}_j(x)f(x)dx = \int_0^\infty \widetilde{F}_j(y)g(y)dy$$

$$> k_{\lambda,j}(\sigma) \left[\int_0^\infty x^{p(1-\mu)-1} f^p(x)dx \right]^{\frac{1}{p}}$$

$$\times \left[\int_0^\infty y^{q(1-\sigma)-1} g^q(y)dy \right]^{\frac{1}{q}}, \tag{6.44}$$

$$\left(\int_0^\infty y^{p\sigma-1} \widetilde{F}_j^p(y)dy \right)^{\frac{1}{p}} > k_{\lambda,j}(\sigma)$$

$$\times \left[\int_0^\infty x^{p(1-\mu)-1} f^p(x)dx \right]^{\frac{1}{p}}, \tag{6.45}$$

$$\left(\int_0^\infty x^{q\mu-1} \widetilde{G}_j^q(x)dx \right)^{\frac{1}{q}} > k_{\lambda,j}(\sigma)$$

$$\times \left[\int_0^\infty y^{q(1-\sigma)-1} g^q(y)dy \right]^{\frac{1}{q}}. \tag{6.46}$$

Corollary 6.23. *For* $0 < p < 1, j = 1,2$, *if there exists a constant* $\delta_0 > 0$ *such that* $k_{\lambda,j}(\sigma \pm \delta_0) < \infty$, *then the following* *statements,* (I), (II), (III), *and* (IV), *are equivalent:*

(I) Both

$$k_{\lambda,j}^{\frac{1}{p}}(\sigma)k_{\lambda,j}^{\frac{1}{q}}(\lambda-\mu) \quad \text{and} \quad k_{\lambda,j}\left(\frac{\sigma}{p}+\frac{\lambda-\mu}{q}\right)$$

are independent of p and q;

(II) $k_{\lambda,j}^{\frac{1}{p}}(\sigma)k_{\lambda}^{\frac{1}{q}}(\lambda-\mu) = k_{\lambda,j}(\frac{\sigma}{p}+\frac{\lambda-\mu}{q})$;

(III) for $\mu+\sigma-\lambda \in (q\delta_0, -q\delta_0)$, we have $\mu+\sigma=\lambda$;

(IV) the constant factor

$$k_{\lambda,j}^{\frac{1}{p}}(\sigma)k_{\lambda,j}^{\frac{1}{q}}(\lambda-\mu)$$

in (6.41)–(6.43) is the best possible.

Example 6.24. (i) For $\sigma > 0, j = 1, h(u) = 1$,

$$k_1(\sigma) = \int_0^1 u^{\sigma-1}du = \frac{1}{\sigma},$$

in (6.32), (6.37), and (6.38), we have the following equivalent Hardy-type integral inequalities with the best possible constant factor $\frac{1}{\sigma}$:

$$\int_0^\infty \left(\int_0^{\frac{1}{y}} f(x)dx\right) g(y)dy = \int_0^\infty \left(\int_0^{\frac{1}{x}} g(y)dy\right) f(x)dx$$

$$> \frac{1}{\sigma}\left[\int_0^\infty x^{p(1-\sigma)-1}f^p(x)dx\right]^{\frac{1}{p}}\left[\int_0^\infty y^{q(1-\sigma)-1}g^q(y)dy\right]^{\frac{1}{q}}, \quad (6.47)$$

$$\left[\int_0^\infty y^{p\sigma-1}\left(\int_0^{\frac{1}{y}} f(x)dx\right)^p dy\right]^{\frac{1}{p}} > \frac{1}{\sigma}\left[\int_0^\infty x^{p(1-\sigma)-1}f^p(x)dx\right]^{\frac{1}{p}},$$

$$(6.48)$$

$$\left[\int_0^\infty x^{q\sigma-1}\left(\int_0^{\frac{1}{x}} g(y)dy\right)^q dx\right]^{\frac{1}{q}} > \frac{1}{\sigma}\left[\int_0^\infty y^{q(1-\sigma)-1}g^q(y)dy\right]^{\frac{1}{q}}.$$

$$(6.49)$$

(ii) For $\sigma < 0, j = 2, h(u) = 1$,

$$k_2(\sigma) = \int_1^\infty u^{\sigma-1}du = \frac{-1}{\sigma},$$

in (6.32), (6.37), and (6.38), replacing σ by $-\sigma$, we have the following equivalent Hardy-type integral inequalities with the best possible constant factor $\frac{1}{\sigma}(\sigma > 0)$:

$$
\int_0^\infty \left(\int_{\frac{1}{y}}^\infty f(x)dx \right) g(y)dy = \int_0^\infty \left(\int_{\frac{1}{x}}^\infty g(y)dy \right) f(x)dx
$$

$$
> \frac{1}{\sigma} \left[\int_0^\infty x^{p(1+\sigma)-1} f^p(x)dx \right]^{\frac{1}{p}} \left[\int_0^\infty y^{q(1+\sigma)-1} g^q(y)dy \right]^{\frac{1}{q}}, \quad (6.50)
$$

$$
\left[\int_0^\infty y^{-p\sigma-1} \left(\int_{\frac{1}{y}}^\infty f(x)dx \right)^p dy \right]^{\frac{1}{p}} > \frac{1}{\sigma} \left[\int_0^\infty x^{p(1+\sigma)-1} f^p(x)dx \right]^{\frac{1}{p}},
$$

$$
(6.51)
$$

$$
\left[\int_0^\infty x^{-q\sigma-1} \left(\int_{\frac{1}{x}}^\infty g(y)dy \right)^q dx \right]^{\frac{1}{q}} > \frac{1}{\sigma} \left[\int_0^\infty y^{q(1+\sigma)-1} g^q(y)dy \right]^{\frac{1}{q}}.
$$

$$
(6.52)
$$

(iii) For $\sigma > 0, j = 1, k_\lambda(1, u) = 1$,

$$
k_{\lambda,1}(\sigma) = \int_0^1 u^{\sigma-1}du = \frac{1}{\sigma},
$$

$$
k_\lambda(x, y) = x^{-\lambda}k_\lambda\left(1, \frac{y}{x}\right) = x^{-\lambda}(\lambda \in \mathbf{R})
$$

in (6.44)–(6.46), we have the following equivalent Hardy-type integral inequalities with the best possible constant factor $\frac{1}{\sigma}(\mu = \lambda - \sigma)$:

$$
\int_0^\infty \left(\int_0^y x^{-\lambda}f(x)dx \right) g(y)dy
$$

$$
= \int_0^\infty \left(\int_0^x g(y)dy \right) x^{-\lambda}f(x)dx
$$

$$
> \frac{1}{\sigma} \left[\int_0^\infty x^{p(1+\sigma-\lambda)-1} f^p(x)dx \right]^{\frac{1}{p}} \left[\int_0^\infty y^{q(1-\sigma)-1} g^q(y)dy \right]^{\frac{1}{q}},
$$

$$
(6.53)
$$

$$\left[\int_0^\infty y^{p\sigma-1}\left(\int_0^y x^{-\lambda}f(x)dx\right)^p dy\right]^{\frac{1}{p}}$$

$$> \frac{1}{\sigma}\left[\int_0^\infty x^{p(1+\sigma-\lambda)-1}f^p(x)dx\right]^{\frac{1}{p}}, \qquad (6.54)$$

$$\left[\int_0^\infty x^{-q\sigma-1}\left(\int_0^x g(y)dy\right)^q dx\right]^{\frac{1}{q}}$$

$$> \frac{1}{\sigma}\left[\int_0^\infty y^{q(1-\sigma)-1}g^q(y)dy\right]^{\frac{1}{q}}. \qquad (6.55)$$

(iv) For $\sigma < 0$, $j = 2$, $k_\lambda(1,u) = 1$,

$$k_{\lambda,2}(\sigma) = \int_1^\infty u^{\sigma-1}du = \frac{-1}{\sigma},$$

$$k_\lambda(x,y) = x^{-\lambda}k_\lambda\left(1,\frac{y}{x}\right) = x^{-\lambda}(\lambda \in \mathbf{R})$$

in (6.44)–(6.46), replacing σ by $-\sigma$ ($\mu = \lambda + \sigma$), we have the following equivalent Hardy-type integral inequalities with the best possible constant factor $\frac{1}{\sigma}(\sigma > 0)$:

$$\int_0^\infty \left(\int_y^\infty x^{-\lambda}f(x)dx\right)g(y)dy = \int_0^\infty \left(\int_x^\infty g(y)dy\right)x^{-\lambda}f(x)dx$$

$$> \frac{1}{\sigma}\left[\int_0^\infty x^{p(1-\sigma-\lambda)-1}f^p(x)dx\right]^{\frac{1}{p}}\left[\int_0^\infty y^{q(1+\sigma)-1}g^q(y)dy\right]^{\frac{1}{q}}, \qquad (6.56)$$

$$\left[\int_0^\infty y^{-p\sigma-1}\left(\int_y^\infty x^{-\lambda}f(x)dx\right)^p dy\right]^{\frac{1}{p}}$$

$$> \frac{1}{\sigma}\left[\int_0^\infty x^{p(1-\sigma-\lambda)-1}f^p(x)dx\right]^{\frac{1}{p}}, \qquad (6.57)$$

$$\left[\int_0^\infty x^{q\sigma-1}\left(\int_x^\infty g(y)dy\right)^q dx\right]^{\frac{1}{q}} > \frac{1}{\sigma}\left[\int_0^\infty y^{q(1+\sigma)-1}g^q(y)dy\right]^{\frac{1}{q}}. \qquad (6.58)$$

References

[1] Schur, I.: Bernerkungen sur theorie der beschrankten bilinearformen mit unendich vielen veranderlichen. *J. Math.*, **140**, 1–28 (1911).

[2] Carleman, T.: Sur les equations integrals singulieres a noyau reel et symetrique. Uppsala (1923).

[3] Wilhelm, M.: On the spectrum of Hilbert's inequality. *Am. J. Math.*, **72**, 699–704 (1950).

[4] Zhang, K.W.: A bilinear inequality. *J. Math. Anal. Appl.*, **271**, 288–296 (2002).

[5] Hardy, G.H.: Note on a theorem of Hilbert concerning series of positive terms. *Proc. London Math. Soc.*, **23**, 45–46 (1925).

[6] Hardy, G.H., Littlewood, J.E., Pólya, G.: *Inequalities*. Cambridge University Press, Cambridge (1934).

[7] Mitrinović, D.S., Pečarić, J.E., Fink, A.M.: *Inequalities Involving Functions and Their Integrals and Derivatives*. Kluwer Academic Publishers, Boston, USA (1991).

[8] Yang, B.C.: On Hilbert's integral inequality. *J. Math. Anal. Appl.*, **220**, 778–785 (1998).

[9] Yang, B.C.: A note on Hilbert's integral inequality. *Chin. Q. J. Math.*, **13**(4), 83–86 (1998).

[10] Yang, B.C.: On an extension of Hilbert's integral inequality with some parameters. *Aust. J. Math. Anal. Appl.*, **1**(1), Art. 11: 1–8 (2004).

[11] Yang, B.C.: *Hilbert-Type Integral Inequalities*. Bentham Science Publishers Ltd., The United Arab Emirates (2009).

[12] Yang, B.C.: *Discrete Hilbert-type Inequalities*. Bentham Science Publishers Ltd., The United Arab Emirates (2011).

[13] Yang, B.C.: A mixed Hilbert-type inequality with a best constant factor. *Int. J. Pure Appl. Math.*, **20**(3), 319–328 (2005).

[14] Yang, B.C., Chen, Q.: A half discrete Hilbert-type inequality with a homogeneous kernel and an extension. *J. Ineq. Appl.*, **124**, 9 (2011). doi:10.1186/1029-242X-2011-124.

[15] Yang, B.C.: A half discrete Hilbert-type inequality with a non-homogeneous kernel and two variables. *Mediterr. J. Math.*, **10**, 677–692 (2013).

[16] Yang, B.C.: Half-discrete Hilbert-type inequalities, operators and compositions. *Handbook of Functional Equations, Functional Inequalities* (Ed. Th. M. Rassias), Springer, New York, pp. 459–534 (2015).

[17] Rassias, M. Th., Yang, B.C.: A half-discrete Hardy-Hilbert-type inequality with a best possible constant factor related to the Hurwitz zeta function. *Progress in Approximation Theory and Applicable Complex Analysis: In Memory of Q.I. Rahman* (Eds. N.K. Govil, R.N. Mohapatra, M.A. Qazi, and G. Schmeisser), Springer, New York, pp. 183–218 (2017).

[18] Yang, B.C., Rassias, Th. M.: On the study of Hilbert-type inequalities with multi-parameters: A survey. *Int. J. Nonlinear Anal. Appl.*, **2**(1), 21–34 (2011).

[19] Debnath, L., Yang, B.C.: Recent developments of Hilbert-type discrete and integral inequalities with applications. *Int. J. Math. Math. Sci.*, **2012**, 29 (2012).

[20] Hong, H.: On multiple Hardy-Hilbert integral inequalities with some parameters. *J. Ineq. Appl.*, **2006**, 11 (2006).

[21] Zhong, W.Y., Yang, B.C.: On a multiple Hilbert-type integral inequality with symmetric kernel. *J. Ineq. Appl.*, **2007**, 17 (2007).

[22] Yang, B.C., Krnić, M.: On the norm of a multidimensional Hilbert-type operator. *Sarajevo J. Math.*, **7**(20), 223–243 (2011).

[23] Krnić, M., Pečarić, J.E., Vuković, P.: On some higher-dimensional Hilbert's and Hardy-Hilbert's type integral inequalities with parameters. *Math. Ineq. Appl.*, **11**, 701–716 (2008).

[24] Yang, B.C.: Hilbert-type integral operators: norms and inequalities. In: *Nonlinear Analysis, Stability, Approximation, and Inequalities* (Eds. P. M. Paralos *et al.*). Springer, New York, pp. 771–859 (2012).

[25] Rassias, M. Th., Yang, B.C.: A multidimensional Hilbert-type integral inequality related to the Riemann zeta function. In: *Applications of Mathematics and Informatics in Science and Engineering* (Ed. N.J. Daras). Springer, New York, pp. 417–433 (2014).

[26] Huang, Z.X., Yang, B.C.: A multidimensional Hilbert-type integral inequality. *J. Ineq. Appl.*, **2015**, 151 (2015).

[27] Liu, T., Yang, B.C., He, L.P.: On a multidimensional Hilbert-type integral inequality with logarithm function. *Math. Ineq. Appl.*, **18**(4), 1219–1234 (2015).

[28] Krnić, M., Vuković, P.: On a multidimensional version of the Hilbert-type inequality. *Anal. Math.*, **38**, 291–303 (2012).

[29] Yang, B.C., Chen, Q.: A multidimensional discrete Hilbert-type inequality. *J. Math. Ineq.*, **8**(2), 267–277 (2014).

[30] Chen, Q. Yang, B.C.: On a more accurate multidimensional Mulholland-type inequality. *J. Ineq. Appl.*, **2014**, 322 (2014).

[31] Yang, B.C.: Multidimensional discrete Hilbert-type inequalities, operator and compositions. In: *Analytic Number Theory, Approximation Theory, and Special Functions* (Eds. G.V. Milovanović *et al.*). Springer, New York, pp. 429–484 (2014).

[32] Rassias, M. Th., Yang, B.C.: A multidimensional half-discrete Hilbert-type inequality and the Riemann zeta function. *Appl. Math. Comput.*, **225**, 263–277 (2013).

[33] Rassias, M. Th., Yang, B.C.: On a multidimensional half-discrete Hilbert-type inequality related to the hyperbolic cotangent function. *Appl. Math. Comput.*, **242**, 800–813 (2014).

[34] Yang, B.C.: On a more accurate multidimensional Hilbert-type inequality with parameters. *Math. Ineq. Appl.*, **18**(2), 429–441 (2015).

[35] Yang, B.C.: On more accurate reverse multidimensional half-discrete Hilbert-type inequalities. *Math. Ineq. Appl.*, **18**(2), 589–605 (2015).

[36] Shi, Y.P., Yang, B.C.: On a multidimensional Hilbert-type inequality with parameters. *J. Ineq. Appl.* **2015**, 371 (2015).

[37] Yang, B.C.: Multidimensional Hilbert-type integral inequalities and their operators expressions. In: *Topics in Mathematical Analysis and Applications* (Eds. Th. M. Rassias and László Tóth), Springer, New York, pp. 769–814 (2015) 3002.

[38] Yang, B.C.: Multidimensional half-discrete Hilbert-type inequalities and operator expressions. *Mathematics Without Boundaries: Surveys in Pure Mathematics* (Eds. Th. M. Rassias and Panos M. Pardalos), Springer, New York, pp. 651–724 (2015).

[39] Zhong, J.H., Yang, B.C.: An extension of a multidimensional Hilbert-type inequality. *J. Ineq. Appl.*, **2017**, 78 (2017).

[40] Yang, B.C., Chen, Q.: A more accurate multidimensional Hardy-Mulholland-type inequality with a general homogeneous kernel. *J. Math. Ineq.*, **12**(1), 113–128 (2018).

[41] Yang, B.C.: A more accurate multidimensional Hardy-Hilbert's inequality. *J. Appl. Anal. Comput.*, **8**(2), 558–572 (2018).

[42] Yang, B.C.: Multidimensional discrete Hilbert-type inequalities, operator and compositions. In: *Analytic Number Theory, Approximation Theory, and Special Functions* (Eds. G. V. Milovanovic, M. Th. Rassias), Springer, New York, pp. 429–484 (2014).

[43] Yang, B.C.: Multidimensional Hilbert-type integral inequalities and their operators expressions. In: *Topics in Mathematical Analysis and Applications* (Eds. Th. M. Rassias and L. Tóth), Springer, New York, pp. 769–814 (2015).

[44] Yang, B.C.: Multidimensional half-discrete Hilbert-type inequalities and operator expressions. In: *Mathematics Without Boundaries: Surveys in Pure Mathematics* (Eds. Th. M. Rassias and P.M. Pardalos), Springer, New York, pp. 651–724 (2015).

[45] Yang, B.C., Chen, Q.: A more accurate multidimensional Hardy–Mulholland-type inequality with a general homogeneous kernel. *J. Math. Ineq.*, **12**(1), 113–128 (2018).

[46] Yang, B.C.: A more accurate multidimensional Hardy–Hilbert's inequality. *J. Appl. Anal. Comput.*, **8**(2), 558–572 (2018).

[47] Yang, B.C.: A more accurate multidimensional Hardy–Hilbert-type inequality. *J. King Saud Univ. Sci.*, **8**(2), 558–572 (2018). doi: 10.1016/j.jksus.2018. 01.004.

[48] Yang, B.C., Liao, J.Q.: *Parameterized Multidimensional Hilbert-type Inequalities*. Scientific Research Publishing, Inc., USA (2020).

[49] Rassias, M. Th., Yang, B.C.: A Hilbert-type integral inequality in the whole plane related to the hyper geometric function and the beta function. *J. Math. Anal. Appl.*, **428**(2), 1286–1308 (2015).

[50] Rassias, M. Th., Yang, B.C.: A more accurate half-discrete Hardy–Hilbert-type inequality with the best possible constant factor related to the extended Riemann-zeta function. *Int. J. Nonlinear Anal. Appl.* **7**(2), 1–27 (2016).

[51] Rassias, M. Th., Yang, B.C.: A half-discrete Hilbert-type inequality in the whole plane related to the Riemann zeta function. *Appl. Anal.*, **97**(9), 1505–1525 (2018). doi: 10.1080/00036811. 2017.1313411.

[52] Rassias, M. Th., Yang, B.C.: Equivalent conditions of a Hardy-type integral inequality related to the extended Riemann zeta function. *Adv. Oper. Theory*, **2**(3), 237–256 (2017).

[53] Liao, J.Q., Yang, B.C.: On Hardy-type integral inequalities with the gamma function. *J. Ineq. Appl.*, **2017**, 131 (2017).

[54] Wang, A.Z., Yang, B.C.: A more accurate half-discrete Hardy–Hilbert-type inequality with the logarithmic function. *J. Ineq. Appl.*, **2017**, 153 (2017).

[55] Rassias, M. Th., Yang, B.C.: Equivalent properties of a Hilbert-type integral inequality with the best constant factor related the Hurwitz zeta function. *Ann. Funct. Anal.*, **9**(2), 282–295 (2018).

[56] Yang, B.C.: Hilbert-type integral operators: Norms and inequalities. In: *Nonlinear Analysis, Stability, Approximation, and Inequalities* (Eds. P.M. Paralos *et al.*), Springer, New York, pp. 771–859 (2012).

[57] Yang, B.C.: Multiple parameterize Yang–Hilbert-type integral inequalities. In: *Computation, Cryptography, and Network Security* (Eds. N. J. Daras, M. Th. Rassias), Springer, New York, pp. 613–633 (2015).

[58] Yang, B.C., Rassias, M. Th.: Parameterized Yang–Hilbert-type integral inequalities and their operator expressions. In: *Computation, Cryptography, and Network Security* (Eds. N. J., M. Th. Rassias), Springer, New York, pp. 635–736 (2015).

[59] Yang, B.C.: Compositional Yang–Hilbert-type integral inequalities and operators. In: *Contributions in Mathematics and Engineering* (Eds. P. M. Pardalos and Th. M. Rassias), Springer, New York, pp. 675–749 (2016).

[60] Rassias, M. Th., Yang, B.C.: On a Hilbert-type integral inequality in the whole plane. In: *Applications of Nonlinear Analysis* (Eds. Th. M. Rassias), Springer, New York, pp. 665–678 (2018).

[61] Yang, B.C.: A multiple Hilbert-type integral inequality in the whole space. In: *Applications of Nonlinear Analysis* (Eds. Th. M. Rassias), Springer, New York, pp. 827–846 (2018).

[62] Rassias, M. Th., Yang, B.C.: On a Hilbert-type integral inequality in the whole plane related to the extended Riemann zeta function. In: *Mathematical Analysis and Applications* (Eds. Th. M. Rassias, P. M. Pardalos), Springer, New York, pp. 511–528 (2019).

[63] Yang, B.C.: Equivalent properties of parameterized Hilbert-type integral inequalities. In: *Mathematical Analysis and Applications* (Eds. Th. M. Rassias, P. M. Pardalos), Springer, New York, pp. 639–676 (2019).

[64] Yang, B.C.: A more accurate Hardy–Hilbert-type inequality with internal variables. In: *Modern Discrete Mathematics and Analysis* (Eds. N. J. Daras, Th. M. Rassias), Springer, New York, pp. 485–504 (2018).

[65] Yang, B.C., Zhong, Y.R., Chen, Q.: On a new extension of Mulholland's inequality in the whole Plane. *J. Funct. Spaces*, **2018**, 8 (2018) 3002.

[66] Wang, A.Z., Yang, B.C.: On a reverse Mulholland's inequality in the whole plane. *J. Ineq. Appl.*, **2018**, 38 (2018).

[67] He, B., Yang, B.C.: A Mulholland-type inequality in the whole plane with multi parameters. *J. King Saud Univ. Sci.*, (2018). http://creativecommons.org/licenses/by-nc-nd/4.0/.

[68] Chen, Q., Yang, B.C.: An extended reverse Hardy–Hilbert's inequality in the whole plane. *J. Ineq. Appl.*, **2018**, 115 (2018).

[69] Rassias, M. Th., Yang, B.C.: On a Hilbert-type integral inequality related to the extended Hurwitz zeta function in the whole plane. *Acta Appl. Math.*, **2018**, 14 (2018). doi: 10.1007/s10440-018-0195-9.

[70] Yang, B.C., Chen, Q.: On a new discrete Mulholland-type inequality in the whole plane. *J. Ineq. Appl.*, **2018**, 184 (2018).

[71] He, L.P., Li, Y., Yang, B.C.: An extended Hilbert's integral inequality in the whole plane with parameters. *J. Ineq. Appl.*, **2018**, 216 (2018).

[72] Zhong, Y.R., Huang, M.F., Yang, B.C.: A Hilbert-type integral inequality in the whole plane related to the kernel of exponent function. *J. Ineq. Appl.*, **2018**, 234 (2018).

[73] Rassias, M. Th., Yang, B.C.: On a Hilbert–type integral inequality in the whole plane related to the extended Riemann zeta function. In: *Complex Analysis and Operator Theory* (2018). doi: 10.1007/s11785-018-0830-5.

[74] Rassias, M. Th., Yang, B.C., Raigorodskii, A.: Two kinds of the reverse Hardy-type integral inequalities with the equivalent forms related to the extended Riemann zeta function. *Appl. Anal. Discrete Math.*, **12**, 273–296 (2018).

[75] Yang, B.C.: An extended multidimensional Hardy–Hilbert-type inequality with a general homogeneous kernel. *Int. J. Nonlinear Anal. Appl.*, **9**(2), 131–143 (2018).

[76] Rassias, M. Th., Yang, B.C.: A reverse Mulholland-type inequality in the whole plane with multi-parameters. *Appl. Anal. Discrete Math.*, **13**, 290–308 (2019).

[77] Yang, B.C., Huang, M.F., Zhong, Y.R.: On a parametric Mulholland-type inequality and applications. *Abstr. Appl. Anal.*, **2019**, 8 (2019). doi: 10.1155/2019/8317029.

[78] He, L.P., Liu, H.Y., Yang, B.C.: On a more accurate reverse Mulholland-type inequality with parameters. *J. Ineq. Appl.*, **2019**, 183 (2019).

[79] Yang, B.C.: A more accurate multidimensional Hardy–Hilbert-type inequality. *J. King Saud Univ. Sci.*, **31**, 164–170 (2019).

[80] He, L.P., Liu, H.Y., Yang, B.C.:. Parametric Mulholland-type inequalities. *J. Appl. Anal. Comput.*, **9**(5), 1973–1986 (2019).

[81] Yang, B.C., Hauang, M.F., Zhong, Y.R.: On an extended Hardy–Hilbert's inequality in the whole plane. *J. Appl. Anal. Comput.*, **9**(6), 2124–2136 (2019).

[82] He, B., Yang, B.C.: A Mulholland-type inequality in the whole plane with multi parameters. *J. King Saud Univ. Sci.*, **32**, 245–250 (2020).

[83] Rassias, M. Th., Yang, B.C., Raigorodskii, A.: On Hardy–type integral inequality in the whole plane related to the extended Hurwitz-zeta fanction. *J. Ineq. Appl.*, **2020**, 94 (2020).

[84] Yang, B.C.: *The Norm of Operator and Hilbert-type Inequalities*. Science Press, Beijing, China (2009).

[85] Yang, B.C., Debnath, L.: *Half-discrete Hilbert-type Inequalities*. World Scientific Publishing Co. Pte. Ltd., Singapore (2014).

[86] Yang, B.C.: *Two Kinds of Multiple Half-discrete Hilbert-type Inequalities*. Lambert Academic Publishing, Deutschland, Germany (2012).

[87] Yang, B.C.: *Topics on Half-discrete Hilbert-type Inequalities*. Lambert Academic Publishing, Deutschland, Germany (2013).

[88] Yang, B.C., Huang Q.L.: *Selection of Hilbert-type Inequalities*. Harbin Institute of Technology Press, China (2018).

[89] Yang, B.C., Rassias, M. Th.: *On Hilbert-type and Hardy-type Integral Inequalities and Applications*. Springer, Switzerland (2019).

[90] Yang, B.C., Liao, J.Q., Agarwal, R.P: *Hilbert-type Inequalities: Operators, Compositions and Extensions*. Scientific Research Publishing, USA (2020).

[91] Hong, Y.: On the structure character of Hilbert's type integral inequality with homogeneous kernel and applications. *J. Jilin Univ. (Science Edition)*, **55**(2), 189–194 (2017).

[92] Hong, Y., Huang, Q.L., Yang, B.C., Liao, J.Q.: The necessary and sufficient conditions for the existence of a kind of Hilbert-type multiple integral inequality with the non-homogeneous kernel and its applications. *J. Ineq. Appl.*, **2017**, 316 (2017).

[93] Rassias, M. Th., Yang, B.C.: Equivalent properties of a Hilbert-type integral inequality with the best constant factor related the Hurwitz zeta function. *Ann. Funct. Anal.*, **9**(2), 282–295 (2018).

[94] Yang, B.C., Chen, Q.: Equivalent conditions of existence of a class of reverse Hardy-type integral inequalities with nonhomogeneous kernel. *J. Jilin Univ. (Science Edition)*, **55**(4), 804–808 (2017).

[95] Yang, B.C.: Equivalent conditions of the existence of Hardy-type and Yang-Hilbert-type integral inequalities with the nonhomogeneous kernel. *J. Guangdong Univ. Educ.*, **37**(3), 5–10 (2017).

[96] Yang, B.C.: On some equivalent conditions related to the bounded property of Yang-Hilbert-type operator. *J. Guangdong Univ. Educ.*, **37**(5), 5–11 (2017).

[97] Yang, Z.M., Yang B.C.: Equivalent conditions of the existence of the reverse Hardy-type integral inequalities with the nonhomogeneous kernel. *J. Guangdong Univ. Educ.*, **37**(5), 28–32 (2017).

[98] Xin, D.M., Yang, B.C., Wang, A.Z.: Equivalent property of a Hilbert-type integral inequality related to the beta function in the whole plane. *J. Funct. Spaces*, **2018**, 8 (2018).

[99] Rassias, M. Th., Yang, B.C.: On an equivalent property of a reverse Hilbert-type integral inequality related to the extended Hurwitz-zeta function. *J. Math. Ineq.*, **13**(2), 315 –334 (2019).

[100] Hong, Y., He, B., Yang, B.C.: Necessary and sufficient conditions for the validity of Hilbert type integral inequalities with a class of quasi-homogeneous kernels and its application in operator theory. *J. Math. Ineq.*, **12**(3), 777–788 (2018).

[101] Huang, Z.X., Yang, B.C.: Equivalent property of a half-discrete Hilbert's inequality with parameters. *J. Ineq. Appl.*, **2018**, 333 (2018).

[102] Liao, J.Q., Hong, Y., Yang, B.C.: Equivalent conditions of a Hilbert-type multiple integral inequality holding. *J. Funct. Spaces*, **2020**, 6 (2020).

[103] Yang, B.C., Huang, M.F., Zhong, Y.R.: Equivalent statements of a more accurate extended Mulholland's inequality with a best possible constant factor. *Math. Ineq. Appl.*, **23**(1), 231–244 (2020).

[104] Mo, H.M., Yang, B.C.: Equivalent properties of a Mulholland-type inequality with a best possible constant factor and parameters. *J. Ineq. Appl.*, **2019**, 123 (2019).

[105] Wang, A.Z., Yang, B.C.: Equivalent property of a more accurate half-discrete Hilbert's inequality. *J. Appl. Anal. Comput.*, **10**(3), 920–934 (2020).

[106] Hong, Y., Liao, J.Q., Yang, B.C., Chen, Q.: A class of Hilbert-type multiple integral inequalities with the kernel of generalized homogeneous function and its applications. *J. Ineq. Appl.*, **2020**, 140 (2020).

[107] Wang, A.Z., Yang, B.C., Chen, Q.: Equivalent properties of a reverse's half-discrete Hilbert's inequality. *J. Ineq. Appl.*, **2019**, 279 (2019).

[108] Krnić, M., Pecarić, J.: Extension of Hilbert's inequality. *J. Math. Anal. Appl.*, **324**(1), 150–160 (2006).

[109] Adiyasuren, V., Batbold, T., Azar, L.E.: A new discrete Hilbert-type inequality involving partial sums. *J. Ineq. Appl.*, **2019**, 127 (2019).

[110] Mo, H.M., Yang, B.C.: On a new Hilbert-type integral inequality involving the upper limit functions. *J. Ineq. Appl.*, **2020**, 5 (2020).

[111] Liao, J.Q., Wu, S.H., Yang, B.C.: On a new half-discrete Hilbert-tipe inequality involving the variable upper limit integral and the partial sum. *Mathematics*, **8**, 229 (2020). doi: 10.3390/math8020229.

[112] Huang, Z.X., Shi, Y.P., Yang, B.C.: On a reverse extended Hardy–Hilbert's inequality. *J. Ineq. Appl.*, **2020**, 68 (2020).

[113] Yang, B.C., Wu, S.H., Wang, A.Z.: A new Hilbert-type inequality with positive homogeneous kernel and its equivalent form. *Symmetric*, **12**, 342 (2020). doi: 10.3390/sym12030342.

[114] Yang, B.C., Wu, S.H., Chen, Q.: On an extended Hardy–Littlewood–Polya's inequality. *AIMS Math.*, **5**(2), 1550–1561 (2020).

[115] Yang, B.C., Wu, S.H., Liao, J.Q.: On a new extended Hardy–Hilbert's inequality with parameters. *Mathematics*, **8**(2020) 73. doi: 10.3390/math8010073.

[116] Yang, B.C., Wu, S.H., Wang, A.Z.: On a reverse half-discrete Hardy–Hilbert's inequality with parameters. *Mathematics*, **7**, 1054 (2019).

[117] Luo, R.C., Yang, B.C.: Parameterized discrete Hilbert-type inequalities with intermediate variables. *J. Ineq. Appl.*, **2019**, 142 (2019).

[118] Huang, X.S., Luo, R.C., Yang, B.C.: On a new extended Half-discrete Hilbert's inequality involving partial sums. *J. Ineq. Appl.*, **2020**, 16 (2020).

[119] Kuang, J.C.: *Real and Functional Analysis* (Continuation) (2nd volume). Higher Education Press, Beijing (2015).

[120] Kuang, J.C.: *Applied Inequalities*. 5th edition. Shangdong Science and Technology Press, Jinan, China (2020).

[121] Faye Hayjin Coyle, Γ.M.: *Calculus Course* (2nd Volume). Higher Education Press, Bingjin (2006).

[122] Zhong, Y.Q.: *Introduction to Complex Functions* (3rd volume). Higher Education Press, Beijing (2003).

[123] Wang, Z.X., Guo, D.R.: *Introduction to Special Functions*. Science Press, Beijing (1979).

[124] Rassias, M. Th.: *Problem-Solving and Selected Topics in Number Theory: In the Spirit of the Mathematical Olympiads*. Springer, New York (2011).

[125] Milovanović, G.V., Rassias, M. Th.: (Eds.), *Analytic Number Theory, Approximation Theory and Special Functions*. Springer, New York (2014).

[126] Gupta, V., Rassias, M. Th.: *Moments of Linear Positive Operators and Approximation*. Springer, New York (2019).

Index

Printed in the United States
by Baker & Taylor Publisher Services